Lecture Notes in Computer Science   10088

Commenced Publication in 1973
Founding and Former Series Editors:
Gerhard Goos, Juris Hartmanis, and Jan van Leeuwen

Editorial Board

David Hutchison
Lancaster University, Lancaster, UK
Takeo Kanade
Carnegie Mellon University, Pittsburgh, PA, USA
Josef Kittler
University of Surrey, Guildford, UK
Jon M. Kleinberg
Cornell University, Ithaca, NY, USA
Friedemann Mattern
ETH Zurich, Zurich, Switzerland
John C. Mitchell
Stanford University, Stanford, CA, USA
Moni Naor
Weizmann Institute of Science, Rehovot, Israel
C. Pandu Rangan
Indian Institute of Technology, Madras, India
Bernhard Steffen
TU Dortmund University, Dortmund, Germany
Demetri Terzopoulos
University of California, Los Angeles, CA, USA
Doug Tygar
University of California, Berkeley, CA, USA
Gerhard Weikum
Max Planck Institute for Informatics, Saarbrücken, Germany

More information about this series at http://www.springer.com/series/7407

Anthony Bonato · Fan Chung Graham
Paweł Prałat (Eds.)

Algorithms and Models for the Web Graph

13th International Workshop, WAW 2016
Montreal, QC, Canada, December 14–15, 2016
Proceedings

 Springer

Editors
Anthony Bonato
Ryerson University
Toronto
Canada

Paweł Prałat
Ryerson University
Toronto
Canada

Fan Chung Graham
University of California San Diego
La Jolla, CA
USA

ISSN 0302-9743   ISSN 1611-3349   (electronic)
Lecture Notes in Computer Science
ISBN 978-3-319-49786-0   ISBN 978-3-319-49787-7   (eBook)
DOI 10.1007/978-3-319-49787-7

Library of Congress Control Number: 2015954997

LNCS Sublibrary: SL1 – Theoretical Computer Science and General Issues

Printed on acid-free paper

This Springer imprint is published by Springer Nature
The registered company is Springer International Publishing AG
The registered company address is: Gewerbestrasse 11, 6330 Cham, Switzerland

Preface

The 13th Workshop on Algorithms and Models for the Web Graph (WAW 2016) took place at the Centre de Recherches Mathématiques, Montréal, Canada, December 14–15, 2016. This is an annual meeting, which is traditionally co-located with another, related, conference. WAW 2016 was co-located with the 12th Conference on Web and Internet Economics (WINE 2016). Co-location of the workshop and conference provided opportunities for researchers in two different but interrelated areas to interact and to exchange research ideas. It was an effective venue for the dissemination of new results and for fostering research collaboration.

The World Wide Web has become part of our everyday life, and information retrieval and data mining on the Web are now of enormous practical interest. The algorithms supporting these activities combine the view of the Web as a text repository and as a graph, induced in various ways by links among pages, hosts, and users. The aim of the workshop was to further the understanding of graphs that arise from the Web and various user activities on the Web, and stimulate the development of high-performance algorithms and applications that exploit these graphs. The workshop gathered together researchers who are working on graph-theoretic and algorithmic aspects of related complex networks, including social networks, citation networks, biological networks, molecular networks, and other networks arising from the Internet.

This volume contains the papers presented during the workshop. Each submission was reviewed by Program Committee members. Papers were submitted and reviewed using the EasyChair online system. The committee members decided to accept 13 papers.

December 2016

Anthony Bonato
Fan Chung Graham
Paweł Prałat

Organization

General Chairs

Andrei Z. Broder — Google Research, USA
Fan Chung Graham — University of California San Diego, USA

Andrei Z. Broder	Google Research, USA
Fan Chung Graham	University of California San Diego, USA

Organizing Committee

Anthony Bonato	Ryerson University, Canada
Fan Chung Graham	University of California, San Diego, USA
Paweł Prałat	Ryerson University, Canada

Sponsoring Institutions

Centre de Recherches Mathématiques
Google
Internet Mathematics
Ryerson University
Yandex

Program Committee

Konstantin Avratchenkov	Inria, France
Paolo Boldi	University of Milan, Italy
Anthony Bonato	Ryerson University, Canada
Milan Bradonjic	Bell Laboratories, USA
Michael Brautbar	Toast, Inc., USA
Fan Chung Graham	UC San Diego, USA
Collin Cooper	King's College London, UK
Andrzej Dudek	Western Michigan University, USA
Alan Frieze	Carnegie Mellon University, USA
David Gleich	Purdue University, USA
Jeannette Janssen	Dalhousie University, Canada
Julia Komjathy	Eindhoven University of Technology, The Netherlands
Ravi Kumar	Google, USA
Silvio Lattanzi	Google, USA
Marc Lelarge	Inria, France
Stefano Leonardi	Sapienza University of Rome, Italy
Nelly Litvak	University of Twente, The Netherlands
Michael Mahoney	UC Berkeley, USA

Contents

An Upper Bound on the Burning Number of Graphs

Max R. Land[1] and Linyuan Lu[2(✉)]

[1] Dutch Fork High School, Irmo, SC 29063, USA
max.ruikang.land@gmail.com
[2] University of South Carolina, Columbia, SC 29208, USA
lu@math.sc.edu

Abstract. The burning number $b(G)$ of a graph G was introduced by Bonato, Janssen, and Roshanbin [Lecture Notes in Computer Science 8882 (2014)] to measure the speed of the spread of contagion in a graph. They proved for any connected graph G of order n, $b(G) \leq 2\lceil \sqrt{n} \rceil - 1$, and conjectured that $b(G) \leq \lceil \sqrt{n} \rceil$. In this paper, we proved $b(G) \leq \lceil \frac{-3+\sqrt{24n+33}}{4} \rceil$, which is roughly $\frac{\sqrt{6}}{2}\sqrt{n}$. We also settled the following conjecture of Bonato-Janssen-Roshanbin: $b(G)b(\bar{G}) \leq n + 4$ provided both G and $\bar{G}$ are connected.

Keywords: Burning number · A-burnable · Graph · Tree

1 Introduction

The burning number of a graph was introduced by Bonato-Janssen-Roshanbin [2,3,10]. It is related to contact processes on graphs such as the Firefighter problem [4,6,7]. In the paper [2,3], Bonato-Janssen-Roshanbin considered a graph process which they called *burning*. At the beginning of the process, all vertices are *unburned*. During each round, one may choose an unburned vertex and change its status to *burned*. At the same time, each of the vertices that are already burned, will remain burned and spread to all of its neighbors and change their status to burned. A graph is called *k-burnable* if it can be burned in at most k steps. The *burning number* of a graph G, denoted by $b(G)$, is the minimum number of rounds necessary to burn all vertices of the graph. For example, $b(K_n) = 2$ for $n \geq 2$, $b(P_4) = 2$, and $b(C_5) = 3$. In the paper [3], they proved $b(P_n) = \lceil n^{1/2} \rceil$. Based on this result, Bonato-Janssen-Roshanbin [3] made the following conjecture.

Conjecture 1: *For any connected graph G of order n, $b(G) \leq \lceil n^{1/2} \rceil$.*

Bonato-Janssen-Roshanbin [2,3] proved $b(G) \leq 2\lceil n^{1/2} \rceil - 1$. The previously known bound is due to Bonato et al. [5]:

$$b(G) \leq \left(\sqrt{\frac{32}{19}} + o(1) \right) \sqrt{n}.$$

In this paper, we improved the upper bound of $b(G)$ as follows.

L. Lu—This author was supported in part by NSF grant DMS-1600811.

A. Bonato et al. (Eds.): WAW 2016, LNCS 10088, pp. 1–8, 2016.
DOI: 10.1007/978-3-319-49787-7_1

Theorem 1. *If G is a connected graph of order n, then*

$$b(G) \leq \left\lceil \frac{-3 + \sqrt{24n + 33}}{4} \right\rceil.$$

In the paper [3], Bonato, Janssen, and Roshanbin also considered Nordhaus-Gaddum Type problems on the burning number. Let $\bar{G}$ be the complement graph of the graph G. In [3], they proved $b(G) + b(\bar{G}) \leq n + 2$ and $b(G)b(\bar{G}) \leq 2n$. Both bounds are tight and are achieved by the complete graph and its complement. When both graphs G and $\bar{G}$ are connected, they proved $b(G) + b(\bar{G}) \leq 3\lceil n^{1/2} \rceil - 1$ and $b(G)b(\bar{G}) \leq n + 6$ for all graph G_n of order $n \geq 6$. The following conjecture has been made in [3]:

Conjecture 2: *If both G and $\bar{G}$ are connected graphs of order n, then $b(G)b(\bar{G}) \leq n + 4$.*

Using Theorem 1, we settled this conjecture positively.

Theorem 2. *If both G and $\bar{G}$ are connected graphs of order n, then*

$$b(G)b(\bar{G}) \leq n + 4.$$

The equality holds if and only if $G = C_5$.

Recently, Mitsche, Pralat, and Roshanbin [8] found some general bounds on the burning number of the Cartesian product and the strong product of graphs. In another paper [9], they determined the burning number of Erdős-Renyi's random graph $G(n, p)$.

The paper is organized as follows. In Sect. 2, we generalize k-burnable to A-burnable, which is a key concept for induction. A tight result is proved for A-burnable trees. The proofs of Theorems 1 and 2 are presented in Sect. 3.

2   Notations and Lemmas

For each positive integer k, let $[k]$ denote the set $\{1, 2, \ldots, k\}$. A graph $G = (V, E)$ consists of a set of vertices V and edges E. The *order* of G, denoted by $|G|$, is the number of vertices in G. A graph G is called *connected* if for any two vertices there is a path connecting them. In this paper, we always assume that G is a connected graph. The *distance* between two vertices u and v, denoted by $d(u, v)$, is the length of a shortest path from u to v in graph G. The *eccentricity* of a vertex v is the maximum distance between v and any other vertex in G. The maximum eccentricity is the *diameter* $D(G)$ while the minimum eccentricity is the *radius* $r(G)$. The *center* of G is the set of vertices of eccentricity equal to the radius.

For any nonnegative integer k and a vertex u, the *k-th closed neighborhood* of u is the set of vertices whose distance from u is at most k, and is denoted by $N_k[u]$. From the definition, a graph G is k-burnable if there is a *burning sequence* $v_1, \ldots, v_k$ of vertices such that

$$V \subseteq \bigcup_{i=1}^{k} N_{k-i}[v_i] \tag{1}$$

$$\forall i, j \in [k] \colon d(v_i, v_j) \geq j - i. \tag{2}$$

The burning number $b(G)$ is the smallest integer k such that G is k-burnable. It has been shown that Condition (2) is redundant for the definition of burning number $b(G)$ (see Lemma 1 of [5]). It is often convenient to rewrite Condition (1) by relabeling the vertices in the burning sequence as follows:

$$V \subseteq \bigcup_{i=1}^{k} N_{i-1}[v_i]. \tag{3}$$

This leads to the following generalization, which is very useful for the purpose of induction. For a set (or multiset) A of k positive integers $a_1, a_2, \ldots, a_k$ (not necessarily all distinct), we say a graph G is A-burnable, if there exist k vertices $v_1, v_2, \ldots, v_k$ such that $G \subseteq \cup_{i=1}^{k} N_{a_i-1}[v_i]$. Under this terminology, the burning number $b(G)$ is the least integer k such that G is $[k]$-burnable.

A *tree* is an acyclic connected graph. For any tree T, it is well-known that the center of T consists of either one vertex or two vertices. If the center of T consists of one vertex, then $D(T) = 2r(T)$; otherwise, $D(T) = 2r(T) - 1$. (See [1].)

A *rooted tree* is a tree with one vertex r designated as the *root*. The *height* of a rooted tree is the eccentricity of the root. In a rooted tree, the *parent* of a vertex is the vertex connected to it on the path to the root. A *child* of a vertex v is a vertex of which v is the parent. A *descendent* of any vertex v is any vertex which is either the child of v or is (recursively) the descendent of any of the children of v. A *leaf* vertex is a vertex with degree 1 but not equal to the root. The *subtree rooted at v* is the induced subgraph on the set of v and its all descendents. The important observation is that if a subtree rooted at v is pruned from the whole tree, the remaining part (if non-empty) is still a tree. This observation is very useful for induction.

A *spanning* tree of graph G is a subtree of G that covers all vertices of G. In the papers [2,3], Bonato, Janssen, and Roshanbin proved

$$b(G) = \min\{b(T) : T \text{ is a spanning subtree of } G\}. \tag{4}$$

Thus, it is sufficient to only consider the burning number $b(T)$ for a tree T.

First, we prove a simple lemma which illustrates the idea of the induction.

Lemma 1. *Let $A = \{a_1, a_2, \ldots, a_k\}$ be a set of k nonnegative integers. If a tree T has order at most $\sum_{i=1}^{k} a_i + \max\{a_i \colon 1 \leq i \leq k\} - 1$, then T is A-burnable.*

Proof. Without loss of generality, we can assume that $a_1 \geq a_2 \geq \cdots \geq a_k$. We will use induction on k. Initial case: $k = 1$, $A = \{a_1\}$. We need to prove that if a tree T has at most $2a_1 - 1$ vertices, then T is A-burnable. Note that

$$r(T) \leq \frac{D(T) + 1}{2} \leq \frac{|T|}{2} \leq a_1 - \frac{1}{2}.$$

Since the radius $r(T)$ is an integer, we must have $r(T) \leq a_1 - 1$. Thus, T is $\{a_1\}$-burnable.

Now we assume the statement holds for any set of $k-1$ integers. For any A of k integers $a_1 \geq a_2 \geq \cdots \geq a_k > 0$ and any tree T with at most $2a_1+a_2+\cdots+a_k-1$, we will prove that T is A-burnable. Pick an arbitrary vertex r as the root of T. Let h be the height of this rooted tree. If $h \leq a_1 - 1$, then $V(T) \subseteq N_{a_1-1}(r)$. So T is $\{a_1\}$-burnable, which implies that T is A-burnable.

Now we assume $h \geq a_1$. Select a leaf vertex u such that $d(r, u) = h$. Let v_k be the vertex on the ru-path such that the distance $d(u, v_k) = a_k - 1$. (This is possible since $h \geq a_1 > a_k - 1$. Let T_1 be the subtree rooted at v_k, and $T_2 := T \setminus T_1$ be the remaining subtree. Notice that $|T_1| \geq a_k$. Thus,

$$
\begin{aligned}
|T_2| &= |T| - |T_1| \\
&\leq 2a_1 + a_2 + \cdots + a_k - 1 - a_k \\
&= 2a_1 + a_2 + \cdots + a_{k-1} - 1.
\end{aligned}
$$

By inductive hypothesis, T_2 is $\{a_1, a_2, \ldots, a_{k-1}\}$-burnable. Hence, there exists $k-1$ vertices $v_1, v_2, \ldots, v_{k-1}$ such that $T_2 \subseteq \cup_{i=1}^{k-1} N_{a_i-1}[v_i]$. Also, notice $T_1 \subseteq N_{a_k-1}[v_k]$. Therefore, $T \subseteq \cup_{i=1}^{k} N_{a_i-1}[v_i]$. The proof of the lemma is finished.

Remark 1. The bound in Lemma 1 is tight.

Proof. Consider the following example: for any positive integer a, let $a_1 = a_2 = \cdots = a_k = a$, i.e. A is a multiset consisting of k a's. Now we will construct a tree T as following. First construct $k + 1$ disjoint paths $P_0, P_1, \ldots, P_k$ with each of order a. Create tree T by connecting one endpoint of $P_1, P_2, \ldots, P_k$ to the same endpoint of P_0 (see figure below).

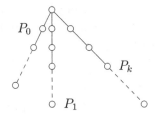

The tree T has order $(k + 1)a$, which is just one more than the amount of vertices in Lemma 1. Now we show T is not A-burnable. Otherwise, there exists $v_1, v_2, \ldots, v_k$ such that T is covered by $\cup_{i=1}^{k} N_{a-1}[v_i]$. By the Pigeon-hole principle, one of the paths $P_0, P_1, \ldots, P_k$ will not contain $v_1, v_2, \ldots, v_k$, and a leaf on this path will not be reached in at most $a - 1$ steps. Thus, T is not A-burnable.

We have the following corollary.

Corollary 1. *For any connected graph G, $b(G) \leq \frac{-3+\sqrt{8n+17}}{2} \approx \sqrt{2n} - \frac{3}{2}$.*

Proof. Let $A = \{k, k-1, \cdots, 1\}$. By Lemma 1, any tree of order $n \leq (\sum_{i=1}^{k} i) + k - 1 = \frac{k^2+3k-2}{2}$ is A-burnable. Solving for k, we get $k \leq \frac{-3+\sqrt{8n+17}}{2}$. Thus, $b(T) \leq \frac{-3+\sqrt{8n+17}}{2}$. By Eq. (4), the same bound holds true for $b(G)$.

3 Proof of Theorems 1 and 2

We have seen that Lemma 1 is sharp when all a_i's are equal. The improvement can be made when a_i's are distinct. We first prove the following lemma.

Lemma 2. *For any $k - 1$ distinct positive integers $a_1 < a_2 < \cdots < a_{k-1}$, there exists an a_i such that $2\lfloor \frac{k-1}{3} \rfloor \leq a_i \leq a_{k-1} - \lfloor \frac{k-1}{3} \rfloor$.*

Proof. Let $j = \lfloor \frac{k-1}{3} \rfloor$ and $A = \{a_1, a_2, \ldots, a_{k-1}\}$. Divide $[1, a_{k-1}]$ into 3 intervals:

$$[1, 2j - 1] \cup [2j, a_{k-1} - j] \cup [a_{k-1} - j + 1, a_{k-1}].$$

There are at most $2j - 1$ elements of A in the first interval. There are at most j elements of A in the last interval. Since $3j - 1 < k - 1$, there exists at least one element of A in the middle interval. Call this element a_i.

Lemma 3. *For all integer $k \geq 1$, we have*

$$\sum_{i=1}^{k} \left\lfloor \frac{i-1}{3} \right\rfloor = \left\lfloor \frac{k^2 - 3k + 2}{6} \right\rfloor.$$

Proof. For $k = 3s$, we have

$$\sum_{i=1}^{k} \left\lfloor \frac{i-1}{3} \right\rfloor = 3 \sum_{j=1}^{s} (j - 1) = \frac{3s(s-1)}{2} = \left\lfloor \frac{k^2 - 3k + 2}{6} \right\rfloor.$$

For $k = 3s + 1$, we have

$$\sum_{i=1}^{k} \left\lfloor \frac{i-1}{3} \right\rfloor = 3 \sum_{j=1}^{s} (j - 1) + s = \frac{3s(s-1)}{2} + s = \left\lfloor \frac{k^2 - 3k + 2}{6} \right\rfloor.$$

For $k = 3s + 2$, we have

$$\sum_{i=1}^{k} \left\lfloor \frac{i-1}{3} \right\rfloor = 3 \sum_{j=1}^{s} (j - 1) + 2s = \frac{3s(s-1)}{2} + 2s = \left\lfloor \frac{k^2 - 3k + 2}{6} \right\rfloor.$$

Theorem 3. *Let A be a set of k distinct positive integers $a_1 < a_2 < \cdots < a_k$. If a tree T has order at most*

$$\left(\sum_{i=1}^{k} a_i \right) + a_k - 1 + \left\lfloor \frac{k^2 - 3k + 2}{6} \right\rfloor,$$

then T is A-burnable.

Proof. Let $f(k) := \lfloor \frac{k^2 - 3k + 2}{6} \rfloor$. By Lemma 3, we have $f(k) = f(k-1) + \lfloor \frac{k-1}{3} \rfloor$. Now, we use induction on k.

Initial case $k = 1$: $A = \{a_1\}$. By Lemma 1, if a tree T has order at most $2a_1 - 1$, then T is $\{a_1\}$-burnable. The statement holds true for $k = 1$ since $f(1) = 0$.

Now assume this statement holds true for any set of $k - 1$ distinct positive integers. Consider the case $A = \{a_1, a_2, \ldots, a_k\}$. We need to prove that if a tree T has order at most $a_1 + a_2 + \cdots + 2a_k - 1 + f(k)$ then T is A-burnable.

Let $j = \lfloor \frac{k-1}{3} \rfloor$. By Lemma 2, there exists a_i that satisfies $2j \le a_i \le a_{k-1} - j$. Choose an arbitrary root r and view T as a rooted tree. Let u be the leaf vertex which has the farthest distance away from the root r. If $d(r, u) \le a_k - 1$, then $V(T) \subseteq N_{a_k - 1}(r)$; thus, T is A-burnable. So, we can assume $d(r, u) \ge a_k$. We will name three vertices v_i, t, v_k on the ru-path such that $d(u, v_i) = a_i - 1$, $d(u, t) = a_i - 1 + j$, and $d(u, v_k) = a_{k-1} - 1$. Let T_1 be the subtree rooted at t. There are two cases:

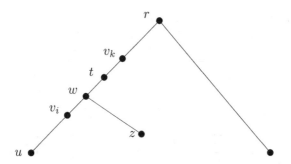

Case 1: $T_1 \subseteq N_{a_i - 1}[v_i]$. Let $T_2 = T \setminus T_1$. Notice $|T_1| \ge a_i + j$. Then,

$$|T_2| = |T| - |T_1|$$
$$\le a_1 + a_2 + \cdots + 2a_k - 1 + f(k) - (a_i + j)$$
$$= a_1 + a_2 + \cdots + a_{i-1} + a_{i+1} + \cdots + 2a_k - 1 + f(k-1).$$

By inductive hypothesis, T_2 is $(A \setminus \{a_i\})$-burnable. Thus, T is A-burnable.

Case 2: $T_1 \not\subseteq N_{a_i - 1}[v_i]$. Then there is a vertex $z \in T_1$, such that $d(v_i, z) \ge a_i$. Let w be the closest vertex on the path rt to z. Observe that w is not in the subtree rooted at v_i. Thus, w is between v_i and t. We have

$$d(w, z) = d(v_i, z) - d(v_i, w) \ge a_i - d(w, v_i) \ge a_i - d(v_i, t) = a_i - j \ge j.$$

The last inequality uses Lemma 2 for the choice of a_i.

Let T_3 be the subtree rooted at v_k and let $T_4 := T \setminus T_3$ be the remaining subtree. We have that $|T_3| \ge a_{k-1} + d(w, z) \ge a_{k-1} + j$. Then,

$$|T_4| = |T| - |T_3|$$
$$\le a_1 + a_2 + \cdots + 2a_k - 1 + f(k) - (a_{k-1} + j)$$
$$= a_1 + a_2 + \cdots + a_{k-2} + 2a_k - 1 + f(k-1).$$

By inductive hypothesis, T_4 is $(A \setminus \{a_{k-1}\})$-burnable. Clearly, T_3 is $\{a_{k-1}\}$-burnable. Therefore, T is A-burnable.

The inductive proof is finished.

Proof (Proof of Theorem 1). Let $A = \{1, 2, \ldots, k\}$. Applying Theorem 3, any tree of n vertices is $[k]$-burnable if

$$n \le 1 + 2 + \cdots + k + k - 1 + \left\lfloor \frac{k^2 - 3k + 2}{6} \right\rfloor = \left\lfloor \frac{2k^2 + 3k - 2}{3} \right\rfloor.$$

Note that $\left\lfloor \frac{2k^2+3k-2}{3} \right\rfloor$ equals to $\frac{2k^2+3k-3}{3}$ if k is divisible by 3; and equals to $\frac{2k^2+3k-2}{3}$ otherwise. In either case, G is $[k]$-burnable if $n \le \frac{2k^2+3k-3}{3}$. Solving for k, we have $k \ge \frac{-3+\sqrt{24n+33}}{4}$. Since k is an integer, we can take ceiling on the bound of k. Thus for any tree T of n vertices,

$$b(T) \le \left\lceil \frac{-3 + \sqrt{24n + 33}}{4} \right\rceil.$$

By Eq. (4), the same bound holds for all connected graphs G.

Lemma 4. *If a graph G is connected and the radius satisfies $r(G) \ge 3$, then the complement graph $\bar{G}$ is also connected and $r(\bar{G}) \le 2$.*

Proof. Since $r(G) \ge 3$, there exists a pair of vertex (u, v) with distance at least 3. Let S be the set of all neighbors of v in the graph G. For any vertex not in $S \cup \{v\}$, it is directly connected to v in the complement graph $\bar{G}$. For any vertex x in S, both xu and uv are edges of $\bar{G}$. Thus, the complement graph $\bar{G}$ has radius at most 2.

Proof (Proof of Theorem 2). By Lemma 4, either $r(G)$ or $r(\bar{G})$ is at most 2. Without loss of generality, we can assume $r(\bar{G}) \le 2$, which implies $b(\bar{G}) \le 3$. We have the following cases.

Case 1: $n \le 4$. Since both G and $\bar{G}$ are connected, the only graph G that can exist is the path P_4. In this case $G = \bar{G} = P_4$. Note, $b(P_4) = 2$. This satisfies

$$b(G) \cdot b(\bar{G}) = 4 < n + 4.$$

Case 2: $n \ge 5$. By Theorem 1, $b(G_n) \le \left\lceil \frac{-3+\sqrt{24n+33}}{4} \right\rceil$.

$$b(G) \cdot b(\bar{G}) \le 3 \cdot \left\lceil \frac{-3 + \sqrt{24n + 33}}{4} \right\rceil.$$

Now we show this bound is at most $n+4$. When $n = 5, 6, 7, 8$, $\left\lceil \frac{-3+\sqrt{24n+33}}{4} \right\rceil = 3$, so $3 \cdot 3 = 9 \le n + 4$. It holds for $n = 5, 6, 7, 8$.

Now we assume $n \geq 9$, we use $\lceil \frac{-3+\sqrt{24n+33}}{4} \rceil \leq \frac{-3+\sqrt{24n+33}}{4} + 1$. It is sufficient to show

$$3\left(\frac{-3 + \sqrt{24n + 33}}{4} + 1 \right) < n + 4.$$

A simple calculation yields $0 < n^2 - 7n - 8$. This true is for all $n \geq 9$.

From the above argument, the equality holds only when $n = 5$ and $b(G) = b(\bar{G}) = 3$. Now assume $n = 5$. If G contains a vertex v of degree 3 or 4, then $b(G) \leq 2$ since $N[v]$ can cover at least 4 vertices. Thus, all degrees of G are at most 2. For the same reason, all degrees of $\bar{G}$ are at most 2. This implies that all degrees in G and in $\bar{G}$ are exactly 2. Since both G and $\bar{G}$ are connected and $n = 5$, the only possible case is $G = \bar{G} = C_5$.

Acknowledgment. We would like to thank the anonymous referees for their helpful comments.

References

1. Biggs, N.L., Lloyd, E.K., Wilson, R.J.: Graph Theory 1736–1936. Oxford University Press, Oxford (1976)
2. Bonato, A., Janssen, J., Roshanbin, E.: Burning a graph as a model of social contagion. In: Bonato, A., Graham, F.C., Prałat, P. (eds.) WAW 2014. LNCS, vol. 8882, pp. 13–22. Springer, Heidelberg (2014). doi:10.1007/978-3-319-13123-8_2
3. Bonato, A., Janssen, J., Roshanbin, E.: How to burn a graph. Internet Math. **12**(1–2), 85–100 (2016)
4. Banerjee, S., Das, A., Gopalan, A., Shakkottai, S.: Epidemic spreading with external agents. IEEE Trans. Inf. Theor. **60**(7), 4125–4138 (2014)
5. Bessy, S., Bonato, A., Janssen, J., Rautenbach, D., Roshanbin, E.: Bounds on the Burning Number. Submitted to Discrete Applied Mathematics
6. Finbow, S., King, A., MacGillivray, G., Rizzi, R.: The firefighter problem for graphs of maximum degree three. Discret. Math. **307**, 2094–2105 (2007)
7. Finbow, S., MacGillivray, G.: The firefighter problem: a survey of results, directions and questions. Australas. J. Comb. **43**, 57–77 (2009)
8. Mitsche, D., Pralat, P., Roshanbin, E.: Burning number of graph products. Submitted to Theoretical Computer Science
9. Mitsche, D., Pralat, P., Roshanbin, E.: Burning graphs—a probabilistic perspective. Submitted to Graphs and Combinatorics
10. Roshanbin, E.: Burning a graph as a model of social contagion. Ph.D. thesis, Dalhousie University (2016)

Assortativity in Generalized Preferential Attachment Models

Alexander Krot[1(✉)] and Liudmila Ostroumova Prokhorenkova[1,2]

[1] Moscow Institute of Physics and Technology, Moscow, Russia
al.krot.kav@gmail.com
[2] Yandex, Moscow, Russia
ostroumova-la@yandex.ru

Abstract. In this paper, we analyze assortativity of preferential attachment models. We deal with a wide class of preferential attachment models (PA-class). It was previously shown that the degree distribution in all models of the PA-class follows a power law. Also, the global and the average local clustering coefficients were analyzed. We expand these results by analyzing the assortativity property of the PA-class of models. Namely, we analyze the behavior of $d_{nn}(d)$ which is the average degree of a neighbor of a vertex of degree d.

Keywords: Networks · Random graphs · Preferential attachment · Assortativity · Average neighbor degree

1 Introduction

Nowadays, there is a great deal of interest in structure and dynamics of real-world networks, from Internet and society networks [1,4,7] to biological networks [2]. The key problem is how to build a model which describes the properties of a given network. Such models are used in physics, information retrieval, data mining, bioinformatics, etc. [1,4,5,17].

Real-world networks have some common properties [12,19,20,22]. For example, for the majority of studied networks, the degree distribution was observed to follow the power law, which means that the portion of vertices with degree d decreases as $d^{-\gamma}$ for some $\gamma > 0$ [3,4,8,11]. Another important property of complex networks is high clustering coefficient [20] which, roughly speaking, measures how likely two neighbors of a vertex are connected.

Another key metric in complex networks analysis is the assortativity coefficient which was first introduced by Newman [18] as the Pearson's correlation coefficient for the pairs $\{(d_i, d_j)|e_{ij} \in E\}$. In assortative graphs edges tend to connect vertices of similar degrees, while in disassortative networks vertices of low degree are more likely to be connected to vertices of high degree. Assortativity coefficient lies between -1 and 1; when this coefficient equals 1, the network is said to have perfect assortative mixing patterns, when it equals 0, the network is non-assortative, while at -1 the network is completely disassortative.

© Springer International Publishing AG 2016
A. Bonato et al. (Eds.): WAW 2016, LNCS 10088, pp. 9–21, 2016.
DOI: 10.1007/978-3-319-49787-7_2

However, as discussed in [13,15], despite Pearson's correlation coefficient is most commonly used to measure assortativity of a network, this coefficient is size-depend when the degree distribution has infinite variance. Another way to analyze assortativity is to consider the behavior of $d_{nn}(d)$ — the average degree of a neighbor of a vertex of degree d. A graph is called assortative if $d_{nn}(d)$ is an increasing function of d, whereas it is referred to as disassortative when $d_{nn}(d)$ is a decreasing function of d. We analyze $d_{nn}(d)$ instead of measuring the correlation since the obtained function of d can give a deeper insight into the network structure.

It was previously shown that in some real-world networks $d_{nn}(d)$ behaves as d^ν for some ν, which can be positive (assortative networks) or negative (disassortative networks) [4,10]. Interestingly, as we show in this paper, in a wide class of preferential attachment models $d_{nn}(d) \propto \log(d)$ as $d \to \infty$.

Assortativity has many applications, for instance, it can be used in the epidemiology. In social networks we usually observe assortative mixing, so diseases targeting high degree individuals are likely to spread to other high degree nodes. On the other hand, biological networks are usually disassortative, therefore vaccination strategies that specifically target the high degree vertices may quickly destroy the epidemic network.

In this paper, we study the behavior of $d_{nn}(d)$ in the T-subclass of the PA-class of models, which was first introduced in [21]. This class includes a lot of well-known models based on the preferential attachment principle: LCD [6], Buckley-Osthus [9], Holme-Kim [14], RAN [23], etc. Despite the fact that the T-subclass generalizes many different models, we are able to analyze $d_{nn}(d)$ in the whole class of models for $\gamma > 3$ (the case of finite variance). We prove that the expectation of $d_{nn}(d)$ asymptotically behaves as $\log(d)$ (up to a constant multiplier). However, this approximation works reasonably well only for very large values of d and for $d < 10^4$ we observe a different behavior which may look like d^ν for some $\nu > 0$.

The remainder of the paper is organized as follows. In Sect. 2, we give a formal definition of the PA-class and present some known results. Then, in Sect. 3, we state new results on the behavior of $d_{nn}(d)$. In Sect. 4, we make some simulations in order to illustrate our results for $d_{nn}(d)$. We prove all theorems in Sect. 5.

2   Generalized Preferential Attachment

2.1   Definition of the PA-Class

Let us formally define the PA-class of models which was first suggested in [21]. Let G_m^n ($n \geq n_0$) be a graph with n vertices $\{1, \ldots, n\}$ and mn edges obtained as a result of the following process. We start at the time n_0 from an arbitrary graph $G_m^{n_0}$ with n_0 vertices and mn_0 edges. On the $(n + 1)$-th step ($n \geq n_0$), we make the graph G_m^{n+1} from G_m^n by adding a new vertex $n + 1$ and m edges connecting this vertex to some m vertices from the set $\{1, \ldots, n, n + 1\}$. Denote by d_v^n the degree of a vertex v in G_m^n. If for some constants A and B the following conditions are satisfied

$$P\left(d_v^{n+1} = d_v^n \mid G_m^n\right) = 1 - A\frac{d_v^n}{n} - B\frac{1}{n} + O\left(\frac{(d_v^n)^2}{n^2}\right), \quad 1 \le v \le n, \qquad (1)$$

$$P\left(d_v^{n+1} = d_v^n + 1 \mid G_m^n\right) = A\frac{d_v^n}{n} + B\frac{1}{n} + O\left(\frac{(d_v^n)^2}{n^2}\right), \quad 1 \le v \le n, \qquad (2)$$

$$P\left(d_v^{n+1} = d_v^n + j \mid G_m^n\right) = O\left(\frac{(d_v^n)^2}{n^2}\right), \quad 2 \le j \le m, \ 1 \le v \le n, \qquad (3)$$

$$P(d_{n+1}^{n+1} = m + j) = O\left(\frac{1}{n}\right), \quad 1 \le j \le m, \qquad (4)$$

then the random graph process G_m^n is a model from the PA-class. Here, as in [21], we require $2mA + B = m$ and $0 \le A \le 1$. We further omit n in d_j^n for simplicity of notation.

As it is explained in [21], even fixing values of parameters A and m does not specify a concrete procedure for constructing a network. There are a lot of models possessing very different properties and satisfying the conditions (1–4), e.g., LCD, Buckley–Osthus, Holme–Kim, and RAN models.

2.2 Power-Law Degree Distribution

Let $N_n(d)$ be the number of vertices of degree d in G_m^n. The following theorems on the expectation of $N_n(d)$ and its concentration were proved in [21].

Theorem 1. *For every model in PA-class and for every* $d = d(n) \ge m$

$$\mathsf{E}N_n(d) = c(m,d)\left(n + O\left(d^{2+\frac{1}{A}}\right)\right),$$

where

$$c(m,d) = \frac{\Gamma\left(d + \frac{B}{A}\right)\Gamma\left(m + \frac{B+1}{A}\right)}{A\Gamma\left(d + \frac{B+A+1}{A}\right)\Gamma\left(m + \frac{B}{A}\right)} \overset{d\to\infty}{\sim} \frac{\Gamma\left(m + \frac{B+1}{A}\right)d^{-1-\frac{1}{A}}}{A\Gamma\left(m + \frac{B}{A}\right)}$$

and $\Gamma(x)$ *is the gamma function.*

Theorem 2. *For every model from the PA-class and for every* $d = d(n)$ *we have*

$$P\left(|N_n(d) - \mathsf{E}N_n(d)| \ge d\sqrt{n}\log n\right) = O\left(n^{-\log n}\right).$$

These two theorems mean that the degree distribution follows (asymptotically) the power law with the parameter $1 + \frac{1}{A}$.

2.3   Clustering Coefficient

A T-subclass of the PA-class was introduced in [21]. In this case, the following additional condition is required:

$$\mathsf{P}\left(d_i^{n+1} = d_i^n + 1, d_j^{n+1} = d_j^n + 1 \mid G_m^n\right) = e_{ij}\frac{D}{mn} + O\left(\frac{d_i^n d_j^n}{n^2}\right), \qquad (5)$$

where $1 \leq i, j \leq n$, e_{ij} is the number of edges between the vertices i and j in G_m^n and D is a non-negative constant. Note that this property still does not define the correlation between edges completely, but it is sufficient for studying the clustering coefficients. Also, this subclass still covers all well-known models mentioned above.

There are two well-known definitions of the clustering coefficient of a graph G. The *global clustering coefficient* $C_1(G)$ is the ratio of three times the number of triangles to the number of pairs of adjacent edges in G. The *average local clustering coefficient* is defined as $C_2(G) = \frac{1}{n}\sum_{i=1}^{n} C(i)$, where $C(i)$ is the local clustering coefficient for a vertex i: $C(i) = \frac{T^i}{P_2^i}$, T^i is the number of edges between the neighbors of the vertex i and P_2^i is the number of pairs of neighbors.

The clustering coefficients for the T-subclass were analyzed in [16,21]. For example, in [21] it was proven that in some cases $(2A \geq 1)$ the global clustering coefficient $C_1(G_m^n)$ tends to zero as the number of vertices grows for all models from the PA-class. Additionally, it was shown that the average local clustering coefficient $C_2(G_m^n)$ does not tend to zero for the T-subclass with $D > 0$. In [16] the local clustering coefficient averaged over the vertices of degree d was analyzed. It was proven that this coefficient $C(d)$ asymptotically decreases as $\frac{2D}{Am} \cdot d^{-1}$ for $A < \frac{3}{4}$.

3   Assortativity

In this paper, we analyze the assortativity property in the T-subclass. One possible way to analyze the assortativity of an undirected graph G is to consider the average degree of the neighbors of vertices with a given degree d:

$$d_{nn}(d) = \frac{1}{N_n(d) \cdot d} \sum_{i:d_i=d} \sum_{j:ij\in E(G)} d_j, \qquad (6)$$

where $E(G)$ is the set of edges of the graph G. If $d_{nn}(d)$ is an increasing function of d, then the network is assortative. Vice-versa, in the disassortative case $d_{nn}(d)$ decreases.

Let $S_n(d)$ be the sum of the degrees of all neighbors of all vertices of degree d. Then $d_{nn}(d)$ can be defined as $d_{nn}(d) = \frac{S_n(d)}{dN_n(d)}$. Hence, in order to estimate $\mathsf{E}d_{nn}(d)$, we first estimate $\mathsf{E}S_n(d)$ and then use Theorems 1 and 2 on the behavior of $N_n(d)$. Namely, we prove the following theorems.

Theorem 3. *Let G_m^n belong to the T-subclass with $A < \frac{1}{2}$. Then, for any $\varepsilon > 0$ and for every $d = d(n) \geq m$*

$$\mathsf{E}S_n(d) = M(d)\left(n + O\left(n^{2A+\varepsilon}d^{2+\frac{1}{A}}\right)\right),$$

where

$$M(d) = (Ad + B + 1)\left[\frac{X}{Am+B+1} + \sum_{i=m+1}^{d} Y(i)\right] \cdot c(m,d),$$

$$X = \frac{m}{A(m-1)+B+1}\left[B - \frac{D}{m} + \frac{(A(m-1)+2B+1)\cdot(Am+B+1)}{1-2A}\right],$$

$$Y(i) = \frac{1}{A(i-1)+B+1}\left[\frac{(B-D/m)i}{Ai+B+1} + \frac{(D/m)\cdot(i-1)}{A(i-1)+B} + m\right].$$

Asymptotically we have

$$M(d) \overset{d\to\infty}{\sim} \frac{Am+B}{A^2} \cdot \frac{\Gamma\left(m+\frac{B+1}{A}\right)}{\Gamma\left(m+\frac{B}{A}\right)} \cdot \log(d) \cdot d^{-\frac{1}{A}}.$$

Theorem 4. *Let G_m^n belong to the T-subclass of the PA-class with $A < \frac{1}{2}$. Then for any $\varepsilon > 0$ and for every $d = d(n) \geq m$*

$$\mathsf{E}d_{nn}(d) = \frac{M(d)}{d\,c(m,d)}\left(1 + O\left(\frac{n^{2A+\varepsilon}d^{2+\frac{1}{A}}}{n} + \frac{d^{2+\frac{1}{A}}\log n}{\sqrt{n}}\right)\right).$$

Note that $\frac{M(d)}{d\cdot c(m,d)} \overset{d\to\infty}{\sim} \frac{Am+B}{A} \cdot \log(d)$.

Note that the restriction $A < \frac{1}{2}$ is essential and for $A \geq \frac{1}{2}$ the result is expected to be completely different. Indeed, when we analyze $S_n(m)$ we have to estimate the expected sum of the degrees of the neighbors of a new vertex i. If $A \geq \frac{1}{2}$, then this sum grows faster than linearly in i and our approximations do not hold.

According to Theorem 4, all networks from the T-subclass with $A < \frac{1}{2}$ are assortative. However, $\mathsf{E}d_{nn}(d)$ increases slowly, as $\log(d)$, unlike d^ν in real-world networks. It is also worth noting that in Theorem 4 we analyze only the average value of $d_{nn}(d)$ and proving concentration is left for future research.

4   Experiments

In this section, we look at the behavior of $d_{nn}(d)$ in a three-parameter model from the family of polynomial graph models defined in [21]. This model belongs to the T-subclass and by varying the parameters we can analyze the effect of A (or, equivalently, γ) on $d_{nn}(d)$. In all the experiments we generated polynomial graphs with $n = 10^6$, $m = 2$, $D = 0.3$ and different values of A. In other words,

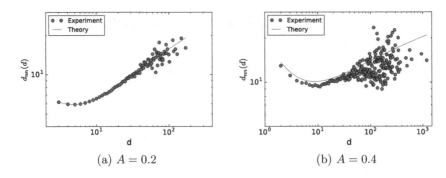

Fig. 1. The behavior of $d_{nn}(d)$ for different A

we fixed the probability of a triangle formation and vary the parameter of the power-law degree distribution. Detailed graph generation process is described in [21].

First, let us illustrate our main result for $\mathsf{E}d_{nn}(d)$ (see Theorem 4). We generated polynomial graphs for different A and compared the obtained values of $d_{nn}(d)$ with their theoretical approximation $\frac{M(d)}{d \cdot c(m,d)}$. We noticed that for $A < \frac{1}{3}$ the theoretical value of $\mathsf{E}d_{nn}(d)$ is extremely close to the experiment. However, if $A > \frac{1}{3}$, then $d_{nn}(d)$ turn out to be consistently smaller than their theoretical approximation. Figure 1 illustrates this observation and shows $d_{nn}(d)$ for $A = 0.2$ and $A = 0.4$. However, according to our additional experiments, the obtained for $A > \frac{1}{3}$ difference tends to zero as $n \to \infty$, as expected. The possible reason for such a slow convergence is the error term $O\left(\frac{n^{3A}}{n^2}\right)$ appearing in the proof in the case $A > \frac{1}{3}$.

We also compared the theoretical value of $\mathsf{E}d_{nn}(d)$ (for $A = 0.2$) with the asymptotic formula $\frac{Am+B}{A} \cdot \log(d)$ (see Fig. 2). Interestingly, from Fig. 1 it may seem that $d_{nn}(d)$ grows as d^{ν} for some ν (as it was observed in real-world networks). However, as d becomes large ($d > 10^4$), one can indeed observe the logarithmic growth.

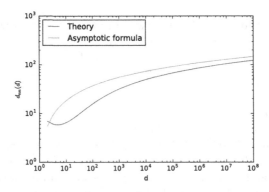

Fig. 2. Theoretical value of $d_{nn}(d)$ versus its asymptotic approximation

5   Proofs

5.1   Proof of Theorem 3

In the proof we use the notation $\theta(\cdot)$ for error terms. By $\theta(X)$ we denote an arbitrary function such that $|\theta(X)| < X$.

We need the following auxiliary theorem.

Theorem 5. *Let W_n be the sum of the squares of the degrees of all vertices in a model from the PA-class with $A < \frac{1}{2}$. Then for any $\epsilon > 0$*

$$\mathsf{E}W_n = \frac{m}{1 - 2A}\,(m + 4B + 1)\,n + O(n^{2A+\epsilon}).$$

Theorem 5 can easily be proved by induction on n. The proof is omitted due to space constraints.

Now let us prove Theorem 3. It can be shown that a.a.s. the maximum degree in G_m^n is less than $n^{A+\varphi}$ for any $\varphi > 0$. Also, $\mathsf{E}N_n(d) = c(m, d)\left(n + O\left(d^{2+\frac{1}{A}}\right)\right)$. Therefore, one can show that $\mathsf{E}S_n(d) = c(m, d)\left(n + O\left(d^{2+\frac{1}{A}}\right)\right)d\,O(n^{A+\varphi})$. As a result, for $n \le Q \cdot d^2$ (for any constant Q) we have $\mathsf{E}S_n(d) = O\left(d^2 n^{A+\varphi}\right) = M(d) \cdot O\left(n^{2A+\varepsilon} \cdot d^{2+\frac{1}{A}}\right)$ for any $\varepsilon > 0$. This concludes the proof for $n \le Qd^2$ for all d.

In order to prove Theorem 3, we use induction on d and for each d we use induction on n. Note that for each d we already have the basis for $n \le Qd^2$.

Consider the case $d = m$. At each step we add a vertex $n + 1$ and m edges. We have the following possibilities.

1. At least one edge hits a vertex of degree m, then $S_n(m)$ is decreased by the sum of the degrees of the neighbors of this vertex. This happens with probability $\frac{A\,m+B}{n} + O\left(\frac{1}{n^2}\right)$. Summing over all vertices of degree m we obtain that $\mathsf{E}S_n(m)$ is decreased by $\left(\frac{Am+B}{n} + O\left(\frac{1}{n^2}\right)\right) \cdot \mathsf{E}S_n(m)$.

2. Exactly one edge hits a neighbor of a vertex of degree m and no edges hit the vertex itself, then $S_n(m)$ is increased by 1. The probability to hit a neighbor is $\frac{Ad_i+B}{n} + O\left(\frac{d_i^2}{n^2}\right)$, where d_i is the degree of this neighbor. We have to subtract the probability to hit both a vertex of degree m and its neighbor which is $\frac{D}{mn} + O\left(\frac{md_i}{n^2}\right)$. Summing over all neighbors of all vertices of degree m, we obtain that $\mathsf{E}S_n(m)$ is increased by:

$$\frac{AES_n(m)}{n} + \frac{B - D/m}{n} mEN_n(m) + O\left(\frac{\mathsf{E}\sum\limits_{\substack{i:i \text{ is a neighbor} \\ \text{of a vertex of degree } m}} d_i^2}{n^2}\right)$$

$$= \frac{AES_n(m)}{n} + \frac{B - D/m}{n} mEN_n(m) + O\left(\frac{\max(n, n^{3A})}{n^2}\right).$$

Here we used the fact that:

$$\mathsf{E}\left(\sum_{\substack{i:i \text{ is a neighbor} \\ \text{of a vertex of degree } m}} d_i^2 \right) \le \mathsf{E}\left(\sum_{i \in V(G_n^m)} d_i^3 \right) = O\left(\max(n, n^{3A})\right).$$

3. If $i > 1$ edges hit a neighbor j of a vertex of degree m, which happens with probability $O\left(\frac{d_j^2}{n^2}\right)$, and no edges hit the vertex itself, then $S_n(m)$ is increased by i. Reasoning as above, we obtain that $\mathsf{E}S_n(m)$ is increased by $O\left(\frac{\max(n, n^{3A})}{n^2}\right)$.

4. The vertex $n + 1$ hits some vertices, so $S_n(m)$ is increased by the sum of the degrees of these vertices. The probability to hit a vertex of degree d_i is $\frac{Ad_i + B}{n} + O\left(\frac{d_i^2}{n^2}\right)$ and after that this vertex will have a degree $d_i + 1$. Summing over i we obtain that $\mathsf{E}S_n(m)$ is increased by:

$$\mathsf{E} \sum_{i \in V(G_n^m)} (d_i + 1)\left(\frac{Ad_i + B}{n} + O\left(\frac{d_i^2}{n^2}\right) \right)$$

$$= \frac{A}{n}\mathsf{E}W_n + (2B + 1)m + O\left(\frac{\max(n, n^{3A})}{n^2}\right).$$

Combining all the cases considered above, we get

$$\mathsf{E}S_{n+1}(m) = \mathsf{E}S_n(m) - \left[\frac{Am + B}{n} + O\left(\frac{1}{n^2}\right)\right]\mathsf{E}S_n(m) + \frac{A\mathsf{E}S_n(m)}{n}$$

$$+ \frac{B - D/m}{n}m\mathsf{E}N_n(m) + \frac{A}{n}\mathsf{E}W_n + (2B + 1)m + O\left(\frac{\max(n, n^{3A})}{n^2}\right).$$

We prove by induction on n that $\mathsf{E}S_n(m) = M(m)\left(n + \theta\left(Cn^{2A+\varepsilon}m^{2+\frac{1}{A}}\right)\right)$ for some constant $C > 0$, where

$$M(m) = \frac{m \cdot c(m, m)}{A(m - 1) + B + 1}\left[B - \frac{D}{m} + \frac{(A(m - 1) + 2B + 1) \cdot (Am + B + 1)}{1 - 2A} \right]. \tag{7}$$

Assume that $\mathsf{E}S_i(m) = M(m)\left(i + \theta\left(Ci^{2A+\varepsilon} \cdot m^{2+\frac{1}{A}}\right)\right)$ for all $i < n + 1$ and let us prove this result for $i = n + 1$:

$$\mathsf{E}S_{n+1}(m) = \left[1 - \frac{A(m - 1) + B}{n} + O\left(\frac{1}{n^2}\right)\right] \cdot M(m)\left(n + \theta\left(Cn^{2A+\varepsilon}m^{2+\frac{1}{A}}\right)\right)$$

$$+ \frac{B - D/m}{n}m \cdot c(m, m)\left[n + O(1)\right] + \frac{A}{n} \cdot \frac{m}{1 - 2A}(m + 4B + 1)n$$

$$+ O(n^{2A-1+\epsilon}) + (2B + 1)m + O\left(\frac{\max(n, n^{3A})}{n^2}\right). \tag{8}$$

Here we use that $\mathsf{E}W_n = \frac{m}{1-2A}(m + 4B + 1)n + O(n^{2A+\epsilon})$ and take $\epsilon < \varepsilon$.

Next, we use (7) and the fact that $c(m, m) = 1/(Am + B + 1)$:

$$\mathsf{ES}_{n+1}(m) = M(m)(n+1) + \left[1 - \frac{A(m-1)+B}{n}\right] M(m)\theta \left(Cn^{2A+\varepsilon}m^{2+\frac{1}{A}}\right)$$
$$+ O\left(Cn^{2A-2+\varepsilon}\right) + O(n^{2A-1+\epsilon}).$$

To complete the proof for $d = m$ we have to show that the obtained error term is not greater than $CM(m)m^{2+\frac{1}{A}}(n+1)^{2A+\varepsilon}$ for some large enough C:

$$CM(m)m^{2+\frac{1}{A}}(n+1)^{2A+\varepsilon} \geq \left[1 - \frac{A(m-1)+B}{n}\right] M(m)Cn^{2A+\varepsilon}m^{2+\frac{1}{A}}$$
$$+ O\left(Cn^{2A-2+\varepsilon}\right) + O(n^{2A-1+\epsilon}).$$

This inequality holds for large enough C. This completes the proof for $d = m$.

Now, consider the case $d > m$, $n > Q\,d^2$. Similarly to the previous case, once we add a vertex $n + 1$ and m edges, we have the following possibilities.

1. At least one edge hits a vertex of degree d. In this case, $\mathsf{ES}_n(d)$ is decreased by $\left(\frac{Ad+B}{n} + O\left(\frac{d^2}{n^2}\right)\right) \cdot \mathsf{ES}_n(d)$.

2. One edge hits a vertex of degree $d - 1$, so $S_n(d)$ is increased by the sum of the degrees of the neighbors of this vertex plus the degree of the new vertex. We get

$$\left(\frac{A(d-1)+B}{n} + O\left(\frac{d^2}{n^2}\right)\right) \cdot (\mathsf{ES}_n(d-1) + m \cdot \mathsf{EN}_n(d-1)).$$

Taking into account the case when, in addition, exactly one edge hits a neighbor of this vertex, we get that $\mathsf{ES}_n(d)$ is additionally increased by:

$$(d-1)\mathsf{EN}_n(d-1) \cdot \frac{D}{mn} + O\left(\frac{(d-1)\mathsf{ES}_n(d-1)}{n^2}\right).$$

3. Exactly one edge hits a neighbor of a vertex of degree d and no edges hit the vertex itself. In this case, $\mathsf{ES}_n(d)$ is increased by:

$$\frac{A\mathsf{ES}_n(d)}{n} + \frac{B - D/m}{n}d\mathsf{EN}_n(d) + O\left(\frac{\max(n, n^{3A})}{n^2}\right) + O\left(\frac{d \cdot \mathsf{ES}_n(d)}{n^2}\right).$$

4. All the cases with multiple edges affect $\mathsf{ES}_n(d)$ by:

$$O\left(\frac{\max(n, n^{3A})}{n^2}\right) + O\left(\frac{d^2}{n^2}\right)\mathsf{ES}_n(d) + O\left(\frac{d^3}{n^2}\right)\mathsf{EN}_n(d). \qquad (9)$$

Combining all the cases considered above, we get

$$\mathsf{ES}_{n+1}(d) = \mathsf{ES}_n(d)\left[1 - \frac{A(d-1)+B}{n}\right] + \frac{A(d-1)+B}{n} \cdot \mathsf{ES}_n(d-1)$$
$$+ \left(\frac{D(d-1)}{mn} + m\frac{A(d-1)+B}{n}\right)\mathsf{EN}_n(d-1) + \frac{(B-D/m)d}{n}\mathsf{EN}_n(d)$$
$$+ O\left(\frac{d^2}{n^2}\right)\mathsf{ES}_n(d) + O\left(\frac{d^3}{n^2}\right)\mathsf{EN}_n(d) + O\left(\frac{\max(n, n^{3A})}{n^2}\right).$$

We prove by induction on d and n that $\mathsf{E}S_n(d) = M(d)\left(n + \theta\left(Cn^{2A+\varepsilon}d^{2+\frac{1}{A}}\right)\right)$ for some constant $C > 0$. Assume that $\mathsf{E}S_i(\tilde{d}) = M(\tilde{d})\left(i + \theta\left(Ci^{2A+\varepsilon}\tilde{d}^{2+\frac{1}{A}}\right)\right)$ for $\tilde{d} < d$ and all i and for $\tilde{d} = d$ and $i < n+1$. Then

$$\mathsf{E}S_{n+1}(d) = \left[1 - \frac{A(d-1)+B}{n}\right]M(d)\left[n + \theta\left(Cn^{2A+\varepsilon}d^{2+\frac{1}{A}}\right)\right]$$
$$+ \frac{A(d-1)+B}{n}M(d-1)\left[n + \theta\left(Cn^{2A+\varepsilon}(d-1)^{2+\frac{1}{A}}\right)\right]$$
$$+ \left(\frac{D(d-1)}{mn} + m\frac{A(d-1)+B}{n}\right)c(m,d-1)\left[n + O\left(d^{2+\frac{1}{A}}\right)\right]$$
$$+ \frac{(B - D/m)d}{n}c(m,d)\left[n + O\left(d^{2+\frac{1}{A}}\right)\right] + O\left(\frac{d^2}{n^2}\right)M(d)\left[n + \theta\left(Cn^{2A+\varepsilon}d^{2+\frac{1}{A}}\right)\right]$$
$$+ O\left(\frac{d^3}{n^2}\right)c(m,d)\left[n + O\left(d^{2+\frac{1}{A}}\right)\right] + O\left(\frac{\max(n, n^{3A})}{n^2}\right).$$

Note that

$$M(d) = \frac{A(d-1)+B}{A(d-1)+B+1}M(d-1) + \frac{(B - D/m)d}{A(d-1)+B+1}c(m,d) \qquad (10)$$
$$+ \frac{\left(\frac{D}{m} + Am\right)(d-1) + Bm}{A(d-1)+B+1}c(m,d-1).$$

Therefore, we obtain:

$$\mathsf{E}S_{n+1}(d) = M(d)(n+1) + \left[1 - \frac{A(d-1)+B}{n}\right]M(d)\,\theta\left(Cn^{2A+\varepsilon}d^{2+\frac{1}{A}}\right)$$
$$+ \frac{A(d-1)+B}{n}M(d-1)\,\theta\left(Cn^{2A+\varepsilon}(d-1)^{2+\frac{1}{A}}\right)$$
$$+ O\left(C\frac{d^4\log(d)\cdot n^{2A+\varepsilon}}{n^2}\right) + O\left(C\frac{d^{2-\frac{1}{A}}\log(d)}{n}\right) + O\left(\frac{\max(n, n^{3A})}{n^2}\right) + O\left(\frac{d^2}{n}\right).$$

It remains to prove that for some large enough C

$$CM(d)\cdot(n+1)^{2A+\varepsilon}d^{2+\frac{1}{A}}$$
$$\geq CM(d)\cdot n^{2A+\varepsilon}d^{2+\frac{1}{A}} - CM(d)\left(\frac{A(d-1)+B}{n}\right)\cdot n^{2A+\varepsilon}d^{2+\frac{1}{A}}$$
$$+ CM(d-1)\left(\frac{A(d-1)+B}{n}\right)\cdot n^{2A+\varepsilon}(d-1)^{2+\frac{1}{A}}$$
$$+ O\left(C\frac{d^4\log(d)\cdot n^{2A+\varepsilon}}{n^2}\right) + O\left(C\frac{d^{2-\frac{1}{A}}\log(d)}{n}\right) + O\left(\frac{\max(n, n^{3A})}{n^2}\right) + O\left(\frac{d^2}{n}\right).$$
$$\tag{11}$$

First, note that

$$CM(d) \cdot (n+1)^{2A+\varepsilon} d^{2+\frac{1}{A}} - CM(d) \cdot n^{2A+\varepsilon} d^{2+\frac{1}{A}}$$

$$= CM(d) \cdot n^{2A+\varepsilon} \cdot d^{2+\frac{1}{A}} \left[\frac{2A+\varepsilon}{n} + O\left(\frac{1}{n^2}\right) \right].$$

Second, one can show that

$$CM(d) \left(\frac{A(d-1)+B}{n} \right) d^{2+\frac{1}{A}} - CM(d-1) \left(\frac{A(d-1)+B}{n} \right) (d-1)^{2+\frac{1}{A}} \geq 0$$

using Eq. (10) and the inequality $(1 - \frac{1}{d})^{-(2+\frac{1}{A})} \geq 1 + \frac{2A+1}{Ad}$.

Therefore, Eq. (11) becomes:

$$CM(d) \cdot n^{2A+\varepsilon} \cdot d^{2+\frac{1}{A}} \left[\frac{2A+\varepsilon}{n} + O\left(\frac{1}{n^2}\right) \right] \geq O\left(C \frac{d^4 \log(d) \cdot n^{2A+\varepsilon}}{n^2} \right)$$

$$+ O\left(C \frac{d^{2-\frac{1}{A}} \log(d)}{n} \right) + O\left(\frac{\max(n, n^{3A})}{n^2} \right) + O\left(\frac{d^2}{n} \right).$$

It is easy to see that for some large enough C and for $n \geq Q \cdot d^2$ (for some large enough Q) this inequality is satisfied. This concludes the proof of the theorem.

5.2 Proof of Theorem 4

Denote by Q the event $\{|N_n(d) - EN_n(d)| < d\sqrt{n}\log(n)\}$. According to Theorem 2, $P(Q) = 1 - O\left(n^{-\log(n)}\right)$. Let us estimate $Ed_{nn}(d)$:

$$Ed_{nn}(d) = E\left(\frac{S_n(d)}{d N_n(d)} \right) = E\left(\frac{S_n(d)}{d N_n(d)} \Big| Q \right) P(Q) + E\left(\frac{S_n(d)}{d N_n(d)} \Big| \bar{Q} \right) P(\bar{Q}).$$

Let us estimate the first term:

$$E\left(\frac{S_n(d)}{d N_n(d)} \Big| Q \right) P(Q) = \frac{E\left(S_n(d) | Q \right) P(Q)}{d\left(EN_n(d) + O\left(d\sqrt{n}\log(n) \right) \right)}$$

$$= \frac{ES_n(d) - E(S_n(d)|\bar{Q}) P(\bar{Q})}{d\left(EN_n(d) + O\left(d\sqrt{n}\log(n) \right) \right)} = \frac{ES_n(d) + O\left(n^{2-\log(n)} \right)}{d\left(EN_n(d) + O\left(d\sqrt{n}\log(n) \right) \right)}.$$

Here we used that $S_n(d) = O(n^2)$. The second term can be estimated as:

$$E\left(\frac{S_n(d)}{d N_n(d)} \Big| \bar{Q} \right) P(\bar{Q}) = O\left(\frac{n^2}{d} \right) P(\bar{Q}) = O\left(\frac{n^{2-\log(n)}}{d} \right).$$

Finally,

$$Ed_{nn}(d) = \frac{M(d)\left(n + O\left(n^{2A+\varepsilon} \cdot d^{2+\frac{1}{A}} \right) \right) + O\left(n^{2-\log(n)} \right)}{d\left(c(m,d)\left(n + O\left(d^{2+\frac{1}{A}} \right) \right) + O\left(d\sqrt{n}\log(n) \right) \right)} + O\left(\frac{n^{2-\log(n)}}{d} \right)$$

$$= \frac{M(d)}{d\, c(m,d)} \left(1 + O\left(\frac{n^{2A+\varepsilon} \cdot d^{2+\frac{1}{A}}}{n} + \frac{d^{2+\frac{1}{A}} \log(n)}{\sqrt{n}} \right) \right).$$

References

1. Albert, R., Barabási, A.-L.: Statistical mechanics of complex networks. Rev. Mod. Phys. **74**, 47–97 (2002)
2. Bansal, S., Khandelwal, S., Meyers, L.A.: Exploring biological network structure with clustered random networks. BMC Bioinform. **10**, 405 (2009)
3. Barabási, A.L., Albert, R.: Emergence of scaling in random networks. Science **286**, 509–512 (1999)
4. Boccaletti, S., Latora, V., Moreno, Y., Chavez, M., Hwang, D.-U.: Complex networks: structure and dynamics. Phys. Rep. **424**(45), 175–308 (2006)
5. Bollobás, B., Riordan, O.M.: Mathematical results on scale-free random graphs. In: Handbook of Graphs, Networks: From the Genome to the Internet, pp. 1–34 (2003)
6. Bollobás, B., Riordan, O.M., Spencer, J., Tusnády, G.: The degree sequence of a scale-free random graph process. Random Struct. Algorithms **18**(3), 279–290 (2001)
7. Borgs, C., Brautbar, M., Chayes, J., Khanna, S., Lucier, B.: The power of local information in social networks. In: Goldberg, P.W. (ed.) WINE 2012. LNCS, vol. 7695, pp. 406–419. Springer, Heidelberg (2012). doi:10.1007/978-3-642-35311-6_30
8. Broder, A., Kumar, R., Maghoul, F., Raghavan, P., Rajagopalan, S., Stata, R., Tomkins, A., Wiener, J.: Graph structure in the web. Comput. Netw. **33**(16), 309–320 (2000)
9. Buckley, P.G., Osthus, D.: Popularity based random graph models leading to a scale-free degree sequence. Discret. Math. **282**, 53–63 (2004)
10. Echenique, P., Gómez-Gardeñes, J., Moreno, Y., Vázquez, A.: Distance-d covering problems in scale-free networks with degree correlations. Phys. Rev. E **71**, 035102(R) (2005)
11. Faloutsos, M., Faloutsos, P., Faloutsos, C.: On power-law relationships of the Internet topology. In: Proceedings of the SIGCOMM 1999 (1999)
12. Girvan, M., Newman, M.E.: Community structure in social and biological networks. Proc. National Acad. Sci. **99**(12), 7821–7826 (2002)
13. van der Hofstad, R., Litvak, N.: Degree-degree dependencies in random graphs with heavy-tailed degrees. Internet Math. **10**(3–4), 287–334 (2014)
14. Holme, P., Kim, B.J.: Growing scale-free networks with tunable clustering. Phys. Rev. E **65**(2), 026107 (2002)
15. van der Hoorn, P., Litvak, N.: Degree-degree dependencies in directed networks with heavy-tailed degrees. Internet Math. **11**(2), 155–178 (2015)
16. Krot, A., Ostroumova Prokhorenkova, L.: Local clustering coefficient in generalized preferential attachment models. In: Gleich, D.F., Komjáthy, J., Litvak, N. (eds.) WAW 2015. LNCS, vol. 9479, pp. 15–28. Springer, Heidelberg (2015). doi:10.1007/978-3-319-26784-5_2
17. Leskovec, J.: Dynamics of Large Networks. ProQuest (2008)
18. Newman, M.E.J.: Assortative mixing in networks. Phys. Rev. Lett. **89**(20), 208701 (2002)
19. Newman, M.E.J.: Power laws, Pareto distributions and Zipf's law. Contemp. Phys. **46**(N5), 323–351 (2005)
20. Newman, M.E.J.: The structure and function of complex networks. SIAM Rev. **45**(2), 167–256 (2003)

21. Ostroumova, L., Ryabchenko, A., Samosvat, E.: Generalized preferential attachment: tunable power-law degree distribution and clustering coefficient. In: Bonato, A., Mitzenmacher, M., Prałat, P. (eds.) WAW 2013. LNCS, vol. 8305, pp. 185–202. Springer, Heidelberg (2013). doi:10.1007/978-3-319-03536-9_15

22. Watts, D.J., Strogatz, S.H.: Collective dynamics of 'small-world' networks. Nature **393**, 440–442 (1998)

23. Zhou, T., Yan, G., Wang, B.-H.: Maximal planar networks with large clustering coefficient and power-law degree distribution. Phys. Rev. E **71**(4), 046141 (2005)

Diclique Clustering in a Directed Random Graph

Mindaugas Bloznelis[1] and Lasse Leskelä[2(✉)]

[1] Vilnius University, Vilnius, Lithuania
[2] Aalto University, Espoo, Finland
lasse.leskela@aalto.fi
http://www.mif.vu.lt/~bloznelis/
http://math.aalto.fi/~lleskela/

Abstract. We discuss a notion of clustering for directed graphs, which describes how likely two followers of a node are to follow a common target. The associated network motifs, called dicliques or bi-fans, have been found to be key structural components in various real-world networks. We introduce a two-mode statistical network model consisting of actors and auxiliary attributes, where an actor i decides to follow an actor j whenever i demands an attribute supplied by j. We show that the digraph admits nontrivial clustering properties of the aforementioned type, as well as power-law indegree and outdegree distributions.

Keywords: Intersection graph · Two-mode network · Affiliation network · Digraph · Diclique · Bi-fan · Complex network

1 Introduction

1.1 Clustering in Directed Networks

Many real networks display a tendency to cluster, that is, to form dense local neighborhoods in a globally sparse graph. In an undirected social network this may be phrased as: *your friends are likely to be friends.* This feature is typically quantified in terms of local and global clustering coefficients measuring how likely two neighbors of a node are neighbors [11,13,14,16]. In directed networks there are many ways to define the concept of clustering, for example by considering the thirteen different ways that a set of three nodes may form a weakly connected directed graph [5].

In this paper we discuss a new type of clustering concept which is motivated by directed online social networks, where a directed link $i \to j$ means that an actor i follows actor j. In such networks a natural way to describe clustering is to say that *your followers are likely to follow common targets.* When the topology of the network is unknown and modeled as a random graph distributed according to a probability measure P, the above statement can be expressed as

$$P(i_2 \to i_4 \mid i_1 \to i_3, i_2 \to i_3, i_1 \to i_4) > P(i_2 \to i_4), \tag{1}$$

where 'you' corresponds to actor i_3. Interestingly, the conditional probability on the left can stay bounded away from zero even for sparse random digraphs [9].

© Springer International Publishing AG 2016
A. Bonato et al. (Eds.): WAW 2016, LNCS 10088, pp. 22–33, 2016.
DOI: 10.1007/978-3-319-49787-7_3

The associated subgraph (Fig. 1) is called a *diclique*. Earlier experimental studies have observed that dicliques (a.k.a. bi-fans) constitute a key structural motif in gene regulation networks [10], citation networks, and several types of online social networks [17].

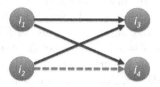

Fig. 1. Forming a diclique by adding a link $i_2 \to i_4$.

Motivated by the above discussion, we define a global *diclique clustering coefficient* of a finite directed graph D with an adjacency matrix (D_{ij}) by

$$\mathcal{C}_{\mathrm{di}}(D) = \frac{\sum_{(i_1,i_2,i_3,i_4)} D_{i_1,i_3} D_{i_1,i_4} D_{i_2,i_3} D_{i_2,i_4}}{\sum_{(i_1,i_2,i_3,i_4)} D_{i_1,i_3} D_{i_1,i_4} D_{i_2,i_3}}, \tag{2}$$

where the sums are computed over all ordered quadruples of distinct nodes. It provides an empirical counterpart to the conditional probability (1) in the sense that the ratio in (2) defines the conditional probability

$$P_D\big(I_2 \to I_4 \mid I_1 \to I_3, I_1 \to I_4, I_2 \to I_3\big), \tag{3}$$

where P_D refers to the distribution of the random quadruple (I_1, I_2, I_3, I_4) sampled uniformly at random among all ordered quadruples of distinct nodes in D.

To quantify diclique clustering among the followers of a selected actor i, we may define a local diclique clustering coefficient by

$$\mathcal{C}_{\mathrm{di}}(D, i) = \frac{\sum_{(i_1,i_2,i_4)} D_{i_1,i} D_{i_1,i_4} D_{i_2,i} D_{i_2,i_4}}{\sum_{(i_1,i_2,i_4)} D_{i_1,i} D_{i_1,i_4} D_{i_2,i}}, \tag{4}$$

where the sums are computed over all ordered triples of distinct nodes excluding i. We remark that $\mathcal{C}_{\mathrm{di}}(D, i) = P_D\big(I_2 \to I_4 \mid I_1 \to I_3, I_1 \to I_4, I_2 \to I_3, I_3 = i\big)$.

Remark 1. By replacing $\to$ by $\leftrightarrow$ in (3), we see that the analogue of the above notion for undirected graphs corresponds to predicting how likely the endpoints of the 3-path $I_2 \leftrightarrow I_3 \leftrightarrow I_1 \leftrightarrow I_4$ are linked together.

1.2 A Directed Random Graph Model

Our goal is to define a parsimonious yet powerful statistical model of a directed social network which displays diclique clustering properties as discussed in the previous section. Clustering properties in many social networks, such as movie actor networks or scientific collaboration networks, are explained by underlying bipartite structures relating actors to movies and scientists to papers [7,12]. Such networks are naturally modeled using directed or undirected random intersection graphs [1,3,4,6,8].

A directed intersection graph on a node set $V = \{1, \ldots, n\}$ is constructed with the help of an auxiliary set of attributes $W = \{w_1, \ldots, w_m\}$ and a directed bipartite graph H with bipartition $V \cup W$, which models how nodes (or actors) relate to attributes. We say that actor i *demands* (or follows) attribute w_k when $i \to w_k$, and *supplies* it when $i \leftarrow w_k$. The directed intersection graph D induced by H is the directed graph on V such that $i \to j$ if and only if H contains a path $i \to w_k \to j$, or equivalently, i demands one or more attributes supplied by j (see Fig. 2). For example, in a citation network the fact that an author i cites a paper w_k coauthored by j, corresponds to $i \to w_k \to j$.

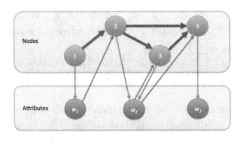

Fig. 2. Node 1 follows node 2, because 1 demands attribute w_1 supplied by 2.

We consider a random bipartite digraph H where the pairs (i, w_k), $i \in V$, $w_k \in W$ establish adjacency relations independently of each other. That is, the bivariate binary random vectors $(\mathbb{I}_{i \to k}, \mathbf{I}_{k \to i})$, $1 \le i \le n$, $1 \le k \le m$, are stochastically independent. Here $\mathbb{I}_{i \to k}$ and $\mathbf{I}_{k \to i}$ stand for the indicators of the events that links $i \to w_k$ and $w_k \to i$ are present in H. We assume that every pair (i, w_k) is assigned a triple of probabilities

$$p_{ik} = P(i \to w_k), \quad q_{ik} = P(w_k \to i), \quad r_{ik} = P(i \to w_k, w_k \to i). \quad (5)$$

Note that, by definition, r_{ik} satisfies the inequalities

$$\max\{p_{ik} + q_{ik} - 1, 0\} \le r_{ik} \le \min\{p_{ik}, q_{ik}\}. \quad (6)$$

A collection of triples $\{(p_{ik}, q_{ik}, r_{ik}), 1 \le i \le n, 1 \le k \le m\}$ defines the distribution of a random bipartite digraph H.

We will focus on a fitness model where every node i is prescribed a pair of weights $x_i, y_i \ge 0$ modelling the demand and supply intensities of i. Similarly, every attribute w_k is prescribed a weight $z_k > 0$ modelling its relative popularity. Denoting $a \wedge b = \min\{a, b\}$ and letting

$$p_{ik} = (\gamma x_i z_k) \wedge 1 \quad \text{and} \quad q_{ik} = (\gamma y_i z_k) \wedge 1 \quad i, k \ge 1, \quad (7)$$

we obtain link probabilities proportional to respective weights. Furthermore, we assume that

$$r_{ik} = r(x_i, y_i, z_k, \gamma), \quad i, k \geq 1, \tag{8}$$

for some function $r \geq 0$ satisfying (6). Here $\gamma > 0$ is a parameter, defining the link density in H, which generally depends on m and n. Note that r defines the correlation between reciprocal links $i \to w_k$ and $w_k \to i$. For example, by letting $r(x, y, z, \gamma) = (\gamma x z \wedge 1)(\gamma y z \wedge 1)$, we obtain a random bipartite digraph with independent links.

We will consider weight sequences having desired statistical properties for complex network modelling. For this purpose we assume that the node and attributes weights are realizations of random sequences $X = (X_i)_{i \geq 1}$, $Y = (Y_i)_{i \geq 1}$, and $Z = (Z_k)_{k \geq 1}$, such that the sequences $\{(X_i, Y_i), i \geq 1\}$ and $\{Z_k, k \geq 1\}$ are mutually independent and consist of independent and identically distributed terms. (However, the attribute weights X_i and Y_i of any given node i are allowed to be correlated with each other.) The resulting random bipartite digraph is denoted by $\mathcal{H}$, and the resulting random intersection digraph by $\mathcal{D}$. We remark that $\mathcal{D}$ extends the random intersection digraph model introduced in [1].

1.3 Degree Distributions

When $\gamma = (mn)^{-1/2}$ and $m, n \to \infty$, the random digraph $\mathcal{D}$ defined in Sect. 1.2 becomes sparse, having the number of links proportional to the number of nodes. Theorem 1 below describes the class of limiting distributions of the outdegree of a typical vertex i. We remark that for each n the outdegrees $d_+(1), \ldots, d_+(n)$ are identically distributed.

To state the theorem, we let $\Lambda_1, \Lambda_2, \Lambda_3$ be mixed-Poisson random variables distributed according to

$$P(\Lambda_i = r) = E\, e^{-\lambda_i} \frac{\lambda_i^r}{r!}, \quad r \geq 0,$$

where $\lambda_1 = X_1 \beta^{1/2} E Z_1$, $\lambda_2 = Z_1 \beta^{-1/2} E Y_1$, and $\lambda_3 = X_1(E Y_1)(E Z_1^2)$. We also denote by Λ_i^* a downshifted size-biased version of Λ_i, distributed according to

$$P(\Lambda_i^* = r) = \frac{r+1}{E \Lambda_i} P(\Lambda_i = r+1), \quad r \geq 0.$$

Below $\xrightarrow{d}$ refers to convergence in distribution.

Theorem 1. *Consider a model with $n, m \to \infty$ and $\gamma = (nm)^{-1/2}$, and assume that $E Y_1, E Z_1^2 < \infty$.*

(i) If $m/n \to 0$ then $d_+(1) \xrightarrow{d} 0$.

(ii) If $m/n \to \beta$ for some $\beta \in (0, \infty)$, then $d_+(1) \xrightarrow{d} \sum_{j=1}^{\Lambda_1} \Lambda_{2,j}^*$, where $\Lambda_{2,1}^*, \Lambda_{2,2}^*, \ldots$ are independent copies of Λ_2^*, and independent of Λ_1.

(iii) If $m/n \to \infty$, then $d_+(1) \xrightarrow{d} \Lambda_3$.

Remark 2. By symmetry, the results of Theorem 1 extend to the indegree $d_-(1)$ when we redefine $\lambda_1 = Y_1 \beta^{1/2} E Z_1$, $\lambda_2 = Z_1 \beta^{-1/2} E X_1$, and $\lambda_3 = Y_1(E X_1)$ $(E Z_1^2)$.

Remark 3. The limiting distributions appearing in Theorem 1: (ii)–(iii) admit heavy tails. This random digraph model is rich enough to model power-law indegree and outdegree distributions, or power-law indegree and light-tailed outdegree distributions [2].

Remark 4. In Theorem 1: (i) it is sufficient to assume that $E Z_1 < \infty$.

We note that a related result for simple (undirected) random intersection graph has been shown in [2]. Theorem 1 extends the result of [2] to digraphs.

1.4   Diclique Clustering

We investigate clustering in the random digraph $\mathcal{D}$ defined in Sect. 1.2 by approximating the (random) diclique clustering coefficient $\mathcal{C}_{\mathrm{di}}(\mathcal{D})$ defined in (2) by a related nonrandom quantity

$$c_{\mathrm{di}} := P(I_2 \to I_4 \mid I_1 \to I_3, \, I_1 \to I_4, \, I_2 \to I_3),$$

where (I_1, I_2, I_3, I_4) is a random ordered quadruple of distinct nodes chosen uniformly at random. Note that here P refers to two independent sources of randomness: the random digraph generation mechanism and the sampling of the nodes. Because the distribution of $\mathcal{D}$ is invariant with respect to a relabeling of the nodes, the above quantity can also be written as

$$c_{\mathrm{di}} = P(2 \to 4 \mid 1 \to 3, \, 1 \to 4, \, 2 \to 3).$$

We believe that under mild regularity conditions $\mathcal{C}_{\mathrm{di}}(\mathcal{D}) \approx c_{\mathrm{di}}$, provided that m and n are sufficiently large. Proving this is left for future work.

Theorem 2 below shows that the random digraph $\mathcal{D}$ admits a *nonvanishing* clustering coefficient c_{di} when the intensity γ is inversely proportional to the number of attributes. For example, by choosing $\gamma = (nm)^{-1/2}$ and letting $m, n \to \infty$ so that $m/n \to \beta > 0$, we obtain a sparse random digraph with tunable clustering coefficient c_{di} and limiting degree distributions defined by Theorem 1 and Remark 2.

Theorem 2. *Assume that $m \to \infty$ and $\gamma m \to \alpha$ for some constant $\alpha \in (0, \infty)$, and that $E X_1^3, E Y_1^3, E Z_1^4 < \infty$. Then*

$$c_{\mathrm{di}} \to \left(1 + \alpha \left(\frac{E X_1^2}{E X_1} + \frac{E Y_1^2}{E Y_1} \right) \frac{(E Z_1^2)(E Z_1^3)}{E Z_1^4} + \alpha^2 \frac{E X_1^2}{E X_1} \frac{E Y_1^2}{E Y_1} \frac{(E Z_1^2)^3}{E Z_1^4} \right)^{-1}.$$

$$(9)$$

Remark 5. When $E X_1^4, E Y_1^4, E Z_1^4 < \infty$, the argument in the proof of Theorem 2 allows to conclude that $c_{\mathrm{di}} \to 0$ when $\gamma m \to \infty$.

To investigate clustering among the followers of a particular node i, we study a theoretical analogue of the local diclique clustering coefficient $\mathcal{C}_{\mathrm{di}}(D, i)$ defined in (4). By symmetry, we may relabel the nodes so that $i = 3$. We will consider the weights of node 3 as known and analyze the conditional probability

$$c_{\mathrm{di}}(X_3, Y_3) = P_{X_3, Y_3}(2 \to 4 \mid 1 \to 3,\ 1 \to 4,\ 2 \to 3),$$

where P_{X_3, Y_3} refers to the conditional probability given (X_3, Y_3). Actually, we may replace P_{X_3, Y_3} by P_{Y_3} above, because all events appearing on the right are independent of X_3.

One may also be interested in analyzing the conditional probability

$$c_{\mathrm{di}}(X, Y) = P_{X, Y}(2 \to 4 \mid 1 \to 3,\ 1 \to 4,\ 2 \to 3),$$

where $P_{X, Y}$ refers to the conditional probability given the values of all node weights $X = (X_i)$ and $Y = (Y_i)$. Again, we may replace $P_{X, Y}$ by P_{X_1, X_2, Y_3, Y_4} above, because the events on the right are independent of the other nodes' weights. More interestingly, $c_{\mathrm{di}}(X, Y)$ turns out to be asymptotically independent of X_2 and Y_4 as well in the sparse regime.

Below $\xrightarrow{P}$ refers to convergence in probability.

Theorem 3. *Assume that $m \to \infty$ and $\gamma m \to \alpha$ for some constant $\alpha \in (0, \infty)$.*

(i) If $E\, X_1^3, E\, Y_1^3, E\, Z_1^4 < \infty$, then

$$c_{\mathrm{di}}(X_3, Y_3) \xrightarrow{P} \left(1 + \alpha \left(\frac{E\, X_1^2}{E\, X_1} + Y_3\right) \frac{(E\, Z_1^3)(E\, Z_1^2)}{E\, Z_1^4} + \alpha^2 Y_3 \frac{E\, X_1^2}{E\, X_1} \frac{(E\, Z_1^2)^3}{E\, Z_1^4}\right)^{-1}.$$

(ii) If $E\, Z_1^4 < \infty$, then

$$c_{\mathrm{di}}(X, Y) \xrightarrow{P} \left(1 + \alpha(X_1 + Y_3)\frac{(E\, Z_1^3)(E\, Z_1^2)}{E\, Z_1^4} + \alpha^2 X_1 Y_3 \frac{(E\, Z_1^2)^3}{E\, Z_1^4}\right)^{-1}.$$

Note that for large Y_3, the clustering coefficient $c_{\mathrm{di}}(X_3, Y_3) = c_{\mathrm{di}}(Y_3)$ scales as Y_3^{-1}. Similarly, for large X_1 and Y_3, the probability $c_{\mathrm{di}}(X, Y)$ scales as $X_1^{-1} Y_3^{-1}$. We remark that similar scaling of a related clustering coefficient in an undirected random intersection graph has been observed in [4].

Remark 6. When all attribute weights are equal to a constant $z > 0$, the statement in Theorem 3: (ii) simplifies into $c_{\mathrm{di}}(X, Y) \xrightarrow{P} (1 + \alpha z X_1)^{-1}(1 + \alpha z Y_3)^{-1}$, a result reported in [9].

Remark 7. Theorems 1, 2, and 3 do not impose any restrictions on the correlation structure of the supply and demand indicators defined by (8).

1.5   Diclique Versus Transitivity Clustering

An interesting question is to compare the diclique clustering coefficient c_{di} with the commonly used transitive closure clustering coefficient

$$c_{\mathrm{tr}} = P\big(2 \to 4 \,|\, 2 \to 3 \to 4\big),$$

see e.g. [5, 15]. The next result illustrates that c_{tr} depends heavily on the correlation between the supply and demand indicators characterized by the function $r(x, y, z, \gamma)$ in (8). A similar finding for a related random intersection graph has been discussed in [1].

Theorem 4. *Let $m, n \to \infty$. Assume that $\gamma = (nm)^{-1/2}$ and $m/n \to \beta$ for some $\beta > 0$. Suppose also that $E\,X_1^2, E\,Y_1^2, E\,Z_1^3 < \infty$.*

(i) If $r(x, y, z, \gamma) = (\gamma xz \wedge 1)(\gamma yz \wedge 1)$, then $c_{\mathrm{tr}} \to 0$.
(ii) If $r(x, y, z, \gamma) = \varepsilon(\gamma xz \wedge \gamma yz \wedge 1)$ for some $0 < \varepsilon \le 1$ and $E\,(X_1 \wedge Y_1) > 0$, then

$$c_{\mathrm{tr}} \; \to \; \left(1 + \frac{\sqrt{\beta}}{\varepsilon}\,\frac{E\,(X_1 Y_1)}{E\,(X_1 \wedge Y_1)}\,\frac{(E\,Z_1^2)^2}{E\,Z_1^3}\right)^{-1}. \tag{10}$$

The assumption in (i) means that the supply and demand indicators of any particular node–attribute pair are conditionally independent given the weights. In contrast, the assumption in (ii) forces a strong correlation between the supply and demand indicators. We note that condition (6) is satisfied in case (ii) for all $i \le n$ and $k \le m$ with high probability as $n, m \to \infty$, because $n^{-1/2} \max_{i \le n}(X_i + Y_i) \xrightarrow{P} 0$ and $m^{-1/2} \max_{k \le m} Z_k \xrightarrow{P} 0$ imply that $\gamma X_i Z_k + \gamma Y_i Z_k \le 1$ for all $i \le n$ and $k \le m$ with high probability.

We remark that in case (i), and in case (ii) with a very small ε, the transitive closure clustering coefficient c_{tr} becomes negligibly small, whereas the diclique clustering coefficient c_{di} remains bounded away from zero. Hence, it makes sense to consider the event $\{1 \to 3, 1 \to 4, 2 \to 3\}$ as a more robust predictor of the link $2 \to 4$ than the event $\{2 \to 3 \to 4\}$. This conclusion has been empirically confirmed for various real-world networks in [10, 17].

2   Proofs

The proof of Theorem 1 goes along similar lines as that of Theorem 1 in [2]. It is omitted. We only give the proofs of Theorems 2 and 3. The proof of Theorem 4 is given in an extended version of the paper available from the authors.

We assume for notational convenience that $\gamma = \alpha m^{-1}$. Denote events $\mathcal{A} = \{1 \to 3, 1 \to 4, 2 \to 3\}$, $\mathcal{B} = \{2 \to 4\}$ and random variables

$$\tilde{p}_{ik} = \alpha \frac{X_i Z_k}{m}, \quad \tilde{q}_{ik} = \alpha \frac{Y_i Z_k}{m}.$$

By $\tilde{P}$ and $\tilde{E}$ we denote the conditional probability and expectation given X, Y, Z. Note that $p_{ik} = \tilde{P}(\mathbb{I}_{i \to k} = 1)$, $q_{ik} = \tilde{P}(\mathbf{I}_{k \to i} = 1)$, and

$$p_{ik} = 1 \wedge \tilde{p}_{ik}, \quad q_{ik} = 1 \wedge \tilde{q}_{ik}. \tag{11}$$

Proof (of Theorem 2). We observe that $\mathcal{A} = \cup_{i \in [4]} \mathcal{A}_i$, where

$$\mathcal{A}_1 = \bigcup_{k \in \mathcal{C}_1} \mathcal{A}_{1.k}, \qquad \mathcal{A}_{1.k} = \{\mathbb{I}_{1 \to k} \mathbb{I}_{2 \to k} \mathbf{I}_{k \to 3} \mathbf{I}_{k \to 4} = 1\},$$

$$\mathcal{A}_2 = \bigcup_{(k,l) \in \mathcal{C}_2} \mathcal{A}_{2.kl}, \qquad \mathcal{A}_{2.kl} = \{\mathbb{I}_{1 \to k} \mathbb{I}_{2 \to l} \mathbf{I}_{k \to 3} \mathbf{I}_{k \to 4} \mathbf{I}_{l \to 3} = 1\},$$

$$\mathcal{A}_3 = \bigcup_{(k,l) \in \mathcal{C}_3} \mathcal{A}_{3.kl}, \qquad \mathcal{A}_{3.kl} = \{\mathbb{I}_{1 \to k} \mathbb{I}_{1 \to l} \mathbb{I}_{2 \to k} \mathbf{I}_{k \to 3} \mathbf{I}_{l \to 4} = 1\},$$

$$\mathcal{A}_4 = \bigcup_{(j,k,l) \in \mathcal{C}_4} \mathcal{A}_{4.jkl}, \qquad \mathcal{A}_{4.jkl} = \{\mathbb{I}_{1 \to j} \mathbb{I}_{1 \to k} \mathbb{I}_{2 \to l} \mathbf{I}_{j \to 3} \mathbf{I}_{k \to 4} \mathbf{I}_{l \to 3} = 1\}.$$

Here $\mathcal{C}_1 = [m]$, $\mathcal{C}_2 = \mathcal{C}_3 = \{(k,l) : k \neq l; \ k, l \in [m]\}$, and $\mathcal{C}_4 = \{(j,k,l) : j \neq k \neq l \neq j; \ j, k, l \in [m]\}$. Hence, by inclusion–exclusion,

$$\sum_{i \in [4]} P(\mathcal{A}_i) - \sum_{\{i,j\} \subset [4]} P(\mathcal{A}_i \cap \mathcal{A}_j) \leq P(\mathcal{A}) \leq \sum_{i \in [4]} P(\mathcal{A}_i).$$

We prove the theorem in Claims 1–3 below. Claim 2 implies that $P(\mathcal{A}) = \sum_{i \in [4]} P(\mathcal{A}_i) + O(m^{-4})$. Claim 3 implies that $P(\mathcal{A} \cap \mathcal{B}) = P(\mathcal{A}_1) + O(m^{-4})$. Finally, Claim 1 establishes the approximation (9) to the ratio $\mathcal{C}_{\mathrm{di}} = P(\mathcal{A} \cap \mathcal{B})/P(\mathcal{A})$.

Claim 1. We have

$$P(\mathcal{A}_1) = \alpha^4 m^{-3} A_1 (1 + o(1)), \tag{12}$$
$$P(\mathcal{A}_2) = \alpha^5 m^{-3} A_2 (1 + o(1)), \tag{13}$$
$$P(\mathcal{A}_3) = \alpha^5 m^{-3} A_3 (1 + o(1)), \tag{14}$$
$$P(\mathcal{A}_4) = \alpha^6 m^{-3} A_4 (1 + o(1)). \tag{15}$$

Here we denote

$$A_1 = a_1^2 b_1^2 h_4, \quad A_2 = a_1^2 b_1 b_2 h_2 h_3, \quad A_3 = a_1 a_2 b_1^2 h_2 h_3, \quad A_4 = a_1 a_2 b_1 b_2 h_2^3.$$

and $a_r = E X_1^r$, $b_r = E Y_1^r$, $h_r = E Z_1^r$.

Claim 2. For $1 \leq i < j \leq 4$ we have

$$P(\mathcal{A}_i \cap \mathcal{A}_j) = O(m^{-4}). \tag{16}$$

Claim 3. We have

$$P(\mathcal{B} \cap \mathcal{A}) = P(\mathcal{A}_1) + O(m^{-4}). \tag{17}$$

Proof of Claim 1. We estimate every $P(\mathcal{A}_r)$ using inclusion-exclusion $I_1 - I_2 \leq P(\mathcal{A}_r) \leq I_1$. Here

$$I_1 = I_1(r) = \sum_{x \in \mathcal{C}_r} P(\mathcal{A}_{r.x}), \qquad I_2 = I_2(r) = \sum_{\{x,y\} \subset \mathcal{C}_r} P(\mathcal{A}_{r.x} \cap \mathcal{A}_{r.y}).$$

Now (12–15) follow from the approximations

$$I_1(1) = \alpha^4 m^{-3} A_1(1 + o(1)), \qquad I_1(2) = \alpha^5 m^{-3} A_2(1 + o(1)), \qquad (18)$$
$$I_1(3) = \alpha^5 m^{-3} A_3(1 + o(1)), \qquad I_1(4) = \alpha^6 m^{-3} A_4(1 + o(1))$$

and bounds $I_2(r) = o(m^{-3})$, for $1 \le r \le 4$.

Firstly we show (18). We only prove the first relation. The remaining cases are treated in much the same way. From the inequalities, see (11),

$$\tilde{p}_{1k}\tilde{p}_{2k}\tilde{q}_{3k}\tilde{q}_{4k} \ge p_{1k}p_{2k}q_{3k}q_{4k} \ge \tilde{p}_{1k}\tilde{p}_{2k}\tilde{q}_{3k}\tilde{q}_{4k}\mathbb{I}'_k \ge \tilde{p}_{1k}\tilde{p}_{2k}\tilde{q}_{3k}\tilde{q}_{4k} - \tilde{p}_{1k}\tilde{p}_{2k}\tilde{q}_{3k}\tilde{q}_{4k}\mathbb{I}^*_k,$$
$$\mathbb{I}'_k = \mathbb{I}_{\tilde{p}_{1k} \le 1}\mathbb{I}_{\tilde{p}_{2k} \le 1}\mathbb{I}_{\tilde{q}_{3k} \le 1}\mathbb{I}_{\tilde{q}_{4k} \le 1}, \qquad \mathbb{I}^*_k = \mathbb{I}_{\tilde{p}_{1k} > 1} + \mathbb{I}_{\tilde{p}_{2k} > 1} + \mathbb{I}_{\tilde{q}_{3k} > 1} + \mathbb{I}_{\tilde{q}_{4k} > 1},$$

we obtain

$$P(\mathcal{A}_{1.k}) = E\, p_{1k}p_{2k}q_{3k}q_{4k} = E\, \tilde{p}_{1k}\tilde{p}_{2k}\tilde{q}_{3k}\tilde{q}_{4k} + R, \qquad (19)$$

where

$$E\, \tilde{p}_{1k}\tilde{p}_{2k}\tilde{q}_{3k}\tilde{q}_{4k} = \alpha^4 m^{-4} A_1 \quad \text{and} \quad |R| \le E\, \tilde{p}_{1k}\tilde{p}_{2k}\tilde{q}_{3k}\tilde{q}_{4k}\mathbb{I}^*_k = o(m^{-4}).$$

The latter limit follows by dominated convergence because $\mathbb{I}^*_k$ is bounded by 4 and tends to zero a.s. as $m \to \infty$. Hence $I_1(1) = mP(\mathcal{A}_{1.k}) = \alpha^4 m^{-3} A_1(1 + o(1))$.

Secondly we show that $I_2(r) = o(m^{-3})$, for $1 \le r \le 4$. For $r = 1$ the bound $I_2(1) = \binom{m}{2} P(\mathcal{A}_{1.k} \cap \mathcal{A}_{1.l}) = o(m^{-3})$ follows from the inequalities

$$P(\mathcal{A}_{1.k} \cap \mathcal{A}_{1.l}) \le E\, \tilde{p}_{1k}\tilde{p}_{2k}\tilde{q}_{3k}\tilde{q}_{4k}\tilde{p}_{1l}\tilde{p}_{2l}\tilde{q}_{3l}\tilde{q}_{4l} = O(m^{-8}).$$

For $r = 2, 3$ we split $I_2(r) = J_1 + \cdots + J_5$, where

$$J_1 = \sum_{\{(k,l),(k,l')\} \subset \mathcal{C}_r} P(\mathcal{A}_{r.kl} \cap \mathcal{A}_{r.kl'}), \qquad J_2 = \sum_{\{(k,l),(k',l)\} \subset \mathcal{C}_r} P(\mathcal{A}_{r.kl} \cap \mathcal{A}_{r.k'l}),$$

$$J_3 = \sum_{\{(k,l),(k',l')\} \subset \mathcal{C}_r} P(\mathcal{A}_{r.kl} \cap \mathcal{A}_{r.k'l'}), \qquad J_4 = \sum_{\{(k,l),(k',k)\} \subset \mathcal{C}_r,\, k' \ne l} P(\mathcal{A}_{r.kl} \cap \mathcal{A}_{r.k'k}),$$

$$J_5 = \sum_{(k,l) \in \mathcal{C}_r} P(\mathcal{A}_{r.kl} \cap \mathcal{A}_{r.lk}).$$

In the first (second) sum distinct pairs $x = (k, l)$ and $y = (k', l')$ share the first (second) coordinate. In the third sum all coordinates of the pairs $(k, l), (k', l')$ are different. In the fourth sum the pairs $(k, l), (k', k)$ only share one common element, but it appears in different coordinates. We show that each $J_i = o(m^{-3})$. Next we only consider the case of $r = 2$. The case of $r = 3$ is treated in a similar way. We have

$$J_1 = m\binom{m-1}{2} P(\mathcal{A}_{2.kl} \cap \mathcal{A}_{2.kl'}) \le m^3 E\, H_1, \quad H_1 = p_{1k}p_{2l}p_{2l'}q_{3k}q_{4k}q_{3l}q_{3l'},$$

$$J_2 = m\binom{m-1}{2} P(\mathcal{A}_{2.kl} \cap \mathcal{A}_{2.k'l}) \le m^3 E\, H_2, \quad H_2 = p_{1k}p_{1k'}p_{2l}q_{3k}q_{4k}q_{3k'}q_{4k'}q_{3l},$$

$$J_3 = \binom{m}{2}\binom{m-2}{2} P(\mathcal{A}_{2.kl} \cap \mathcal{A}_{2.k'l'}) \le m^4 E\, H_3,$$

$$H_3 = p_{1k}p_{1k'}p_{2l}p_{2l'}q_{3k}q_{4k}q_{3k'}q_{4k'}q_{3l}q_{3l'},$$

$$J_4 = m(m-1)(m-2)P(\mathcal{A}_{2.kl} \cap \mathcal{A}_{2.k'k}) \le m^3 E\, H_4,$$

$$H_4 = p_{1k}p_{1k'}p_{2l}p_{2k}q_{3k}q_{4k}q_{3k'}q_{4k'}q_{3l},$$

$$J_5 = \binom{m}{2}P(\mathcal{A}_{2.kl} \cap \mathcal{A}_{2.lk}) \le m^2 E\, H_5, \quad H_5 = p_{1k}p_{1l}p_{2l}p_{2k}q_{3k}q_{3l}q_{4k}q_{4l}.$$

For H_1 we estimate the typical factors $p_{ij} \le \tilde{p}_{ij}$ and $q_{ij} \le \tilde{q}_{ij}$, but

$$q_{3l} \le \tilde{q}_{3l}\mathbb{I}_{Y_3 \le \sqrt{m}} + \mathbb{I}_{Y_3 > \sqrt{m}} \le \alpha m^{-1/2}Z_l + \mathbb{I}_{Y_3 > \sqrt{m}}. \tag{20}$$

We obtain

$$E\, H_1 \le \alpha^6 m^{-6} a_1 a_2 b_1 h_2 h_3 \left(b_2 h_2 \alpha m^{-1/2} + h_1 E\, Y_3^2 \mathbb{I}_{Y_3 > \sqrt{m}}\right) = o(m^{-6}). \tag{21}$$

Hence $J_1 = o(m^{-3})$. Similarly, we show that $J_2 = o(m^{-4})$. Furthermore, while estimating H_3 we apply (20) to q_{3l} and $q_{3l'}$ and apply $p_{ij} \le \tilde{p}_{ij}$ and $q_{ij} \le \tilde{q}_{ij}$ to remaining factors. We obtain

$$H_3 \le \tilde{p}_{1k}\tilde{p}_{1k'}\tilde{p}_{2l}\tilde{p}_{2l'}\tilde{q}_{3k}\tilde{q}_{4k}\tilde{q}_{3k'}\tilde{q}_{4k'}(\alpha m^{-1/2}Z_l + \mathbb{I}_{Y_3 > \sqrt{m}})(\alpha m^{-1/2}Z_{l'} + \mathbb{I}_{Y_3 > \sqrt{m}}). \tag{22}$$

Since the expected value of the product on the right is $o(m^{-8})$, we conclude that $E\, H_3 = o(m^{-8})$. Hence $J_3 = o(m^{-4})$. Proceeding in a similar way we establish the bounds $J_4 = o(m^{-5})$ and $J_5 = O(m^{-6})$.

We explain the truncation step (20) in some more detail. A simple upper bound for H_1 is the product

$$\tilde{p}_{1k}\tilde{p}_{2l}\tilde{p}_{2l'}\tilde{q}_{3k}\tilde{q}_{4k}\tilde{q}_{3l}\tilde{q}_{3l'} = \alpha^7 m^{-7}X_1 X_2^2 Y_3^3 Y_4 Z_k^3 Z_l^2 Z_{l'}^2.$$

It contains an undesirable high power Y_3^3. Using (20) instead of the simple upper bound $q_{3l} \le \tilde{q}_{3l}$ we have reduced in (21) the power of Y_3 down to 2. Similarly, in (22) we have reduced the power of Y_3 from 4 to 2.

Using the truncation argument we obtain the upper bound $I_2(4) = o(m^{-3})$ under moment conditions $E\, X_1^3, E\, Y_1^3, E\, Z_1^4 < \infty$. The proof is similar to that of the bound $I_2(2) = o(m^{-3})$ above. We omit routine, but tedious calculation.

Proof of Claim 2. We only prove that $q := P(\mathcal{A}_3 \cap \mathcal{A}_4) = O(m^{-4})$. The remaining cases are treated in a similar way. For $x = (j,k,l) \in \mathcal{C}_4$ and $y = (r,t) \in \mathcal{C}_3$ we denote, for short, $\mathbb{I}_{\mathcal{A}_{4.x}} = \mathbb{I}_x^* = \mathbb{I}_{jkl}^*$ and $\mathbb{I}_{\mathcal{A}_{3.y}} = \mathbb{I}_y = \mathbb{I}_{rt}$. For $q = E\, \mathbb{I}_{\mathcal{A}_4}\mathbb{I}_{\mathcal{A}_3}$, we write, by symmetry,

$$q \le E\left(\sum_{x \in \mathcal{C}_4} \mathbb{I}_x^*\right)\mathbb{I}_{\mathcal{A}_3} = m(m-1)(m-2)E\, \mathbb{I}_{123}^*\mathbb{I}_{\mathcal{A}_3}$$

and

$$E\, \mathbb{I}_{123}^*\mathbb{I}_{\mathcal{A}_3} \le E\, \mathbb{I}_{123}^*\left(\sum_{y \in \mathcal{C}_3} \mathbb{I}_y\right) = E\, \mathbb{I}_{123}^*(J_1 + J_2 + J_3).$$

Here

$$J_1 = \sum_{r,t\in[m]\setminus[3],\, r\neq t} \mathbb{I}_{rt}, \quad J_2 = \sum_{r\in[m]\setminus[3]} \sum_{s\in[3]} (\mathbb{I}_{sr} + \mathbb{I}_{rs}), \quad J_3 = \sum_{r,t\in[3],\, r\neq t} \mathbb{I}_{rt}.$$

Finally, we show that $E\,\mathbb{I}_{123}^* J_i = O(m^{-7})$, $i \in [3]$. For $i = 1$ we have, by symmetry,

$$E\,\mathbb{I}_{123}^* J_1 = (m-3)(m-4) E\,\mathbb{I}_{123}^* \mathbb{I}_{45}. \tag{23}$$

Invoking the inequalities

$$E\,\mathbb{I}_{123}^* \mathbb{I}_{45} = E\,\tilde{E}\,\mathbb{I}_{123}^* \mathbb{I}_{45} \leq E\,\tilde{p}_{11}\tilde{p}_{12}\tilde{p}_{15}\tilde{p}_{23}\tilde{p}_{24}\tilde{q}_{31}\tilde{q}_{42}\tilde{q}_{33}\tilde{q}_{34}\tilde{q}_{45} = O(m^{-10}) \tag{24}$$

we obtain $E\,\mathbb{I}_{123}^* J_1 = O(m^{-8})$.

The bound $E\,\mathbb{I}_{123}^* J_2 = O(m^{-7})$ is obtained from the identity (which follows by symmetry)

$$E\,\mathbb{I}_{123}^* J_2 = (m-3) \sum_{s\in[3]} \left(E\,\mathbb{I}_{123}^* \mathbb{I}_{s4} + E\,\mathbb{I}_{123}^* \mathbb{I}_{4s} \right),$$

combined with the bounds $E\,\mathbb{I}_{123}^* \mathbb{I}_{s4} + E\,\mathbb{I}_{123}^* \mathbb{I}_{4s} = O(m^{-8})$, $s \in [3]$. We only show the latter bound for $s = 3$. The cases $s = 1,2$ are treated in a similar way. We have

$$E\,\mathbb{I}_{123}^* \mathbb{I}_{34} \leq E\,\tilde{p}_{11}\tilde{p}_{12}\tilde{p}_{13}\tilde{p}_{23}\tilde{q}_{31}\tilde{q}_{42}\tilde{q}_{33}\tilde{q}_{44} = O(m^{-8}),$$

$$E\,\mathbb{I}_{123}^* \mathbb{I}_{43} \leq E\,\tilde{p}_{11}\tilde{p}_{12}\tilde{p}_{13}\tilde{p}_{23}\tilde{p}_{24}\tilde{q}_{31}\tilde{q}_{42}\tilde{q}_{33}\tilde{q}_{34}\tilde{q}_{43} = O(m^{-10}).$$

The proof of $E\,\mathbb{I}_{123}^* J_3 = O(m^{-7})$ is similar. It is omitted.

Proof of Claim 3. We use the notation $\overline{\mathbb{I}}_{\mathcal{A}_j} = 1 - \mathbb{I}_{\mathcal{A}_j}$ for the indicator of the event $\overline{\mathcal{A}}_j$ complement to $\mathcal{A}_j$. For $2 \leq i \leq 4$ we denote $\mathcal{H}_i = (\mathcal{A}_i \cap \mathcal{B}) \setminus \cup_{1\leq j\leq i-1}\mathcal{A}_j$. We have

$$P(\mathcal{A} \cap \mathcal{B}) = P(\cup_{i\in[4]}\mathcal{A}_i \cap \mathcal{B}) = P(\mathcal{A}_1 \cap \mathcal{B}) + R, \qquad 0 \leq R \leq P(\cup_{2\leq i\leq 4}\mathcal{H}_i).$$

Note that $P(\mathcal{A}_1 \cap \mathcal{B}) = P(\mathcal{A}_1)$. It remains to show that $P(\mathcal{H}_i) = O(m^{-4})$, $2 \leq i \leq 4$.

We have, by symmetry,

$$P(\mathcal{H}_2) = E\,\mathbb{I}_{\mathcal{A}_2}\mathbb{I}_{\mathcal{B}}\overline{\mathbb{I}}_{\mathcal{A}_1} \leq E \sum_{x\in\mathcal{C}_2} \mathbb{I}_{\mathcal{A}_{2.x}}\mathbb{I}_{\mathcal{B}}\overline{\mathbb{I}}_{\mathcal{A}_1} = m(m-1)E\,\mathbb{I}_{\mathcal{A}_{2.12}}\mathbb{I}_{\mathcal{B}}\overline{\mathbb{I}}_{\mathcal{A}_1}. \tag{25}$$

Furthermore, we have $\mathbb{I}_{\mathcal{A}_{2.12}}\mathbb{I}_{\mathcal{B}}\overline{\mathbb{I}}_{\mathcal{A}_1} \leq \mathbb{I}_{\mathcal{A}_{2.12}}\left(\mathbb{I}_{2\to 4} + \sum_{3\leq j\leq m} \mathbb{I}_{2\to j}\mathbb{I}_{j\to 4}\right)$ and, by symmetry,

$$E\,\mathbb{I}_{\mathcal{A}_{2.12}}\mathbb{I}_{\mathcal{B}}\overline{\mathbb{I}}_{\mathcal{A}_1} \leq E\,\mathbb{I}_{\mathcal{A}_{2.12}}\mathbb{I}_{2\to 4} + (m-2)E\,\mathbb{I}_{\mathcal{A}_{2.12}}\mathbb{I}_{2\to 3}\mathbb{I}_{3\to 4}.$$

A simple calculation shows that $E\,\mathbb{I}_{\mathcal{A}_{2.12}}\mathbb{I}_{2\to 4} \leq E\,\tilde{p}_{11}\tilde{p}_{22}\tilde{q}_{31}\tilde{q}_{41}\tilde{q}_{32}\tilde{q}_{42} = O(m^{-6})$. Similarly, $E\,\mathbb{I}_{\mathcal{A}_{2.12}}\mathbb{I}_{2\to 3}\mathbb{I}_{3\to 4} = O(m^{-7})$. Therefore, $E\,\mathbb{I}_{\mathcal{A}_{2.12}}\mathbb{I}_{\mathcal{B}}\overline{\mathbb{I}}_{\mathcal{A}_1} = O(m^{-6})$. Now (25) implies $P(\mathcal{H}_2) = O(m^{-4})$. The bounds $P(\mathcal{H}_j) = O(m^{-4})$, $j = 3,4$ are obtained in a similar way.

Proof (of Theorem 3). The proof is the same as that of Theorem 2, but while evaluating the probabilities of events $\mathcal{A}$ and $\mathcal{A} \cap \mathcal{B}$ we treat X_1, X_2, Y_3, Y_4 as constants.

References

1. Bloznelis, M.: A random intersection digraph: indegree and outdegree distributions. Discret. Math. **310**(19), 2560–2566 (2010). http://dx.doi.org/10.1016/j.disc.2010.06.018
2. Bloznelis, M., Damarackas, J.: Degree distribution of an inhomogeneous random intersection graph. Electron. J. Comb. **20**(3), P3 (2013)
3. Bloznelis, M., Godehardt, E., Jaworski, J., Kurauskas, V., Rybarczyk, K.: Recent progress in complex network analysis: properties of random intersection graphs. In: Lausen, B., Krolak-Schwerdt, S., Böhmer, M. (eds.) Data Science, Learning by Latent Structures, and Knowledge Discovery. Studies in Classification, Data Analysis, and Knowledge Organization, pp. 79–88. Springer, Heidelberg (2015). http://dx.doi.org/10.1007/978-3-662-44983-7_7
4. Deijfen, M., Kets, W.: Random intersection graphs with tunable degree distribution and clustering. Probab. Eng. Inf. Sci. **23**(4), 661–674 (2009). http://dx.doi.org/10.1017/S0269964809990064
5. Fagiolo, G.: Clustering in complex directed networks. Phys. Rev. E **76**, 026107 (2007). http://link.aps.org/doi/10.1103/PhysRevE.76.026107
6. Frieze, A., Karoński, M.: Introduction to Random Graphs. Cambridge University Press, Cambridge (2016)
7. Guillaume, J.L., Latapy, M.: Bipartite structure of all complex networks. Inf. Process. Lett. **90**(5), 215–221 (2004)
8. Karoński, M., Scheinerman, E.R., Singer-Cohen, K.B.: On random intersection graphs: the subgraph problem. Comb. Probab. Comput. **8**(1–2), 131–159 (1999). http://dx.doi.org/10.1017/S0963548398003459
9. Leskelä, L.: Directed random intersection graphs. In: 18th INFORMS Applied Probability Society Conference, Istanbul, Turkey, July 2015
10. Milo, R., Shen-Orr, S., Itzkovitz, S., Kashtan, N., Chklovskii, D., Alon, U.: Network motifs: simple building blocks of complex networks. Science **298**(5594), 824–827 (2002). http://science.sciencemag.org/content/298/5594/824
11. Newman, M.E.J.: The structure and function of complex networks. SIAM Rev. **45**(2), 167–256 (2003). http://dx.doi.org/10.1137/S003614450342480
12. Newman, M.E.J., Strogatz, S.H., Watts, D.J.: Random graphs with arbitrary degree distributions and their applications. Phys. Rev. E **64**, 026118 (2001)
13. Scott, J.: Social Network Analysis. SAGE Publications, Thousand Oaks (2012)
14. Szabó, G., Alava, M., Kertész, J.: Clustering in complex networks. In: Ben-Naim, E., Frauenfelder, H., Toroczkai, Z. (eds.) Complex Networks, vol. 650, pp. 139–162. Springer, Heidelberg (2004). http://dx.doi.org/10.1007/978-3-540-44485-5_7
15. Wasserman, S., Faust, K.: Social Network Analysis: Methods and Applications. Cambridge University Press, Cambridge (1994)
16. Watts, D.J., Strogatz, S.H.: Collective dynamics of 'small-world' networks. Nature **393**, 440–442 (1998)
17. Zhang, Q.M., Lü, L., Wang, W.Q., Zhu, Y.X., Tao, Z.: Potential theory for directed networks. PLoS ONE **8**(2), e55437 (2013)

Distributed and Asynchronous Methods
for Semi-supervised Learning

Konstantin Avrachenkov[1]([⊠]), Vivek S. Borkar[2], and Krishnakant Saboo[2]

[1] Inria Sophia Antipolis, Sophia-Antipolis, France
`k.avrachenkov@inria.fr`
[2] IIT Bombay, Mumbai, India
`borkar.vs@gmail.com, kishansaboo.2004@gmail.com`

Abstract. We propose two asynchronously distributed approaches for graph-based semi-supervised learning. The first approach is based on stochastic approximation, whereas the second approach is based on randomized Kaczmarz algorithm. In addition to the possibility of distributed implementation, both approaches can be naturally applied online to streaming data. We analyse both approaches theoretically and by experiments. It appears that there is no clear winner and we provide indications about cases of superiority for each approach.

1 Introduction

Semi-supervised learning (SSL) is a type of learning which uses both labelled and unlabelled data for training [8]. In many practical cases, the amount of labelled data is much less compared to the unlabelled data, making SSL a powerful tool for processing massive data. The present work focuses on graph based SSL [8,11,19,23,28,29]. Consider a (weighted) graph in which nodes belong to one of K classes and the true class of a few nodes is given. An edge and its weight in the graph indicate a similarity and similarity degree between two nodes, respectively. Hence such a graph is called a similarity graph. SSL aims to estimate the class of each of the unlabelled nodes by using the true class information of few labelled nodes and the structure of the graph.

Being of practical importance [11,28], graph based SSL has been widely studied. Most though not all (see e.g., [13,20]) methods formulate SSL as a quadratic optimization problem [2–4,25–27] for feature vectors and then solve it or the associated stationary point equation iteratively. Iterative solutions become computationally intensive because they involve matrix operations which grow in size as a polynomial with the graph size, though the growth can be linear or quasi-linear in case of sparse graphs. The data can be distributed over a network (as in sensor networks [18] or in the internet of things) or can be stored and processed in distributed manner as is the case in data centers. These motivate us to look for distributed algorithms. There are a variety of algorithms that address this issue, see, e.g., the classic text of [5] or a more recent comparative survey of [14]. We apply variants of two of these for the solution of the stationary point equation for

© Springer International Publishing AG 2016
A. Bonato et al. (Eds.): WAW 2016, LNCS 10088, pp. 34–46, 2016.
DOI: 10.1007/978-3-319-49787-7_4

the feature vectors and conduct a comparative analysis. The first, more original, is based on gossip-type stochastic approximation, whereas the second is based on the well known randomized Kaczmarz algorithm. The randomized Kaczmarz algorithm has drawn much attention in recent years, beginning with the seminal work of [22] - see [16,17,30] for some subsequent work ([30] is the closest in spirit to ours). Both approaches are asynchronously distributed and can also be applied online to streaming data (e.g., as in stream image classification tasks [24]). There is no clear winner with respect to accuracy and we provide indications about cases of superiority for each approach. With respect to computational efficiency, the stochastic approximation approach appears to be generally more efficient than the randomized Kaczmarz approach on all our numerical examples. In the case of online data processing, both approaches are simpler and computationally lighter than the method of [24] based on the Nyström approach which is not distributed. We would also like to mention that the recent well performing SSL methods [19,23] use the Jacobi method which can be straightforwardly distributed in synchronous manner. However, we would like to note that the convergence conditions for the asynchronous version of the Jacobi method can be more stringent than the convergence conditions for our approaches [5].

As shall be seen later, our only requirements are: global knowledge on the number of classes and the ability of nodes to pass information to their immediate neighbors. The adjacency matrix of the network is defined by the support of the similarity matrix. Clearly, our approach will be particularly efficient when the similarity matrix is sparse. Regarding mapping data points to agents, our basic scenario is: each node corresponds to a different agent and needs to compute the elements of the classification functions only corresponding to itself. For instance, this is a natural setting in wireless sensor networks. Our approaches can be extended in a straightforward way to the case when one agent is responsible for several nodes. The computation at each node happens in an asynchronous fashion and the information needs to be exchanged only among the neighbours. This highlights the distributed nature of the approaches.

Rest of the article is organised as follows: the next section introduces the problem formally. Section 3 introduces and analyses two distributed approaches for graph-based SSL. Section 4 shows that a number of established SSL methods can be mapped to our general methodology. Section 5 investigates and compares the two approaches by numerical experiments.

2 Definitions and Problem Formulation

Consider a graph with (weighted) adjacency matrix A with N nodes, each belonging to one of K classes. (We discuss online formulation of the problem later.) The class information is given for some nodes, referred to as labelled nodes. Define D as a diagonal matrix with $D_{ii} = d(i)$ for $d(i) :=$ the (weighted) degree of node i. We also use the standard graph Laplacian $L = D - A$. Define $N \times K$ matrix Y containing information about the labelled nodes by:

$$Y_{ij} = \begin{cases} 1 & \text{if labelled node } i \text{ belongs to class } j, \\ 0 & \text{otherwise.} \end{cases}$$

$F = \{F_{ik}\}$ is a $N \times K$ matrix with F_{ik} representing the 'belongingness' of node i to class k. The vector F_{*k} is referred to as 'classification function' or 'feature vector'. The aim of the SSL problem is to find F_{*k} such that it is close to the labelling function and it varies smoothly over the graph. The class of node i is calculated from F by assigning class k to node i if $F_{ik} > F_{ij}, \forall \, j \neq k$. The optimization problem associated with the above stated requirements is to minimize

$$Q(F) = \sum_{k=1}^{K} F_{*k}^{T} \mathcal{A} F_{*k} + \mu \sum_{k=1}^{K} (F_{*k} - Y_{*k})^{T} \mathcal{B} (F_{*k} - Y_{*k}), \tag{1}$$

where $\mathcal{A}$ is the positive (semi-)definite graph kernel and $\mathcal{B}$ the cost of deviation from the labels. Typically, the support of matrix $\mathcal{A}$ coincides with the support of the adjacency matrix A and matrix $\mathcal{B}$ is diagonal. $\mu > 0$ is the regularization parameter. Majority of all existing graph-based semi-supervised learning methods can be cast into the optimization formulation (1). A few important examples are: Choosing $\mathcal{A} = D^{\sigma-1} L D^{\sigma-1}$ and $\mathcal{B} = D^{2\sigma-1}$ [2], we obtain semi-supervised methods based on standard Laplacian ($\sigma = 1$) [26], normalized Laplacian ($\sigma = 1/2$) [25], and PageRank ($\sigma = 0$) [1]. If $\mathcal{A} = L$ and $\mathcal{B} = I$, we retrieve the semi-supervised method based on the regularized Laplacian [3,9,21], which can be viewed as a Lagrangian relaxation of the method based on harmonic functions [27]. If $\mathcal{A} = \gamma D - A$ and $\mathcal{B} = I$, we obtain the method based on the modified regularized Laplacian [15]. A good comparative overview of the graph-based semi-supervised learning methods can be found in [11]. Nearly all methods described in [11] can be represented in optimization formulation (1). In [4,19,23] the authors discussed slightly different quadratic optimization formulations of the semi-supervised learning methods with an additional quadratic extra term. That additional term can be included in the first or second terms of (1).

Except for the harmonic functions method which puts zeros for diagonal elements of $\mathcal{B}$ when labels are unavailable, we used zero as default label when unavailable, a choice neutral to the classes. This gives a bias towards zero for learned classification functions, but because we are interested only in their ordinal comparison, the classification is robust to this choice.

The above problem is a convex quadratic optimization problem. Applying the first order optimality condition and solving for F_{*k} gives the stationary point equation:

$$\left[(\mathcal{A} + \mathcal{A}^T) + \mu (\mathcal{B} + \mathcal{B}^T) \right] F_{*k} = \mu (\mathcal{B} + \mathcal{B}^T) Y_{*k}. \tag{2}$$

3   Distributed Approaches

In this section we describe two approaches for solution of (2) which can be naturally distributed (in asynchronous fashion) and can also be applied in a scenario with streaming data. We prove the convergence of the proposed approaches and comment on their rate of convergence.

3.1 Stochastic Approximation Approach

The first solution is based on Stochastic Approximation (SA-approach). Consider the general problem of finding a unique solution x^* of the system

$$x = G(x) = \widetilde{B}x + \widetilde{Y}, \tag{3}$$

where $x, \widetilde{Y} \in \mathcal{R}^d$ and $\widetilde{B} = \{b(i,j)\} \in \mathcal{R}^{d \times d}$ is irreducible non-negative. In Sect. 4 we show that most semi-supervised learning methods can be written in this form for appropriate $\widetilde{B}$ and $\widetilde{Y}$; see, e.g., (7), (9). Define H as a diagonal matrix with elements as row sums of $\widetilde{B}$, i.e., $H_{ii} = \sum_j \widetilde{B}_{ij}$ Define P as $P = H^{-1}\widetilde{B}$, viewed as the transition probability matrix on the graph, and Q is its irreducible counterpart as in the PageRank algorithm: $Q = (1 - \epsilon)P + \epsilon/N\,E$, where E is an $N \times N$ matrix with all 1's. Let $X_t, t \geq 0$, be a Markov chain with transition matrix Q and $\{\eta_t\}_{t\geq 0}$ a positive step-size sequence satisfying $\sum_{t\geq 0} \eta_t = \infty$ and $\sum_{t\geq 0} \eta_t^2 < \infty$. The stochastic approximation scheme to solve (3) is:

$$x_i^{t+1} = x_i^t + \eta_t I\{X_t = i\} \frac{P(i, X_{t+1})}{Q(i, X_{t+1})} \left(H_{ii} x_{X_{t+1}}^t - x_i^t + \widetilde{Y}_i \right). \tag{4}$$

Convergence Analysis. In Eq. (3), if $\widetilde{B}$ has its Perron-Frobenius eigenvalue $\lambda \in (0,1)$ with the normalized positive eigenvector $w = [w_1, ... w_d]^T$, then the following hold. (Proofs will appear in the journal version of the paper.) Define the weighted norm $\|x\|_w = \max_i |x_j/w_i|$.

Lemma. Map G is a contraction w.r.t. $\|x\|_w$, i.e., $\|G(x) - G(y)\|_w \leq \lambda\|x - y\|_w$.

Using the above lemma, we can establish the following.

Theorem. Almost surely, $x^t \to x^*$ as $t \to \infty$.

Convergence Rate. Section 4.2 of [6] gives results on sample complexity of a synchronous stochastic approximation. The result broadly implies that, at time n, the probability of remaining within a prescribed small neighborhood of x^* after $n + \tau$ iterates is greater than $1 - O\left(e^{-\frac{C}{\sum_{t\geq n} \eta_t^2}}\right)$, $C > 0$. For our case where, $\eta_t = \Theta\left(\frac{1}{t}\right)$, the decay of probability of ever escaping from this neighborhood after $n + \tau$ iterations is exponential. Here τ is a quantity specified in terms of problem parameters.

We used Markov chain sampling of nodes above. For more general asynchronous schemes, the following considerations apply. We can use the step size $\eta_{\nu(t,i)}$ for the ith node where $\nu(t,i) = $ the number of updates node i has made till time t i.e., its local clock. Also, η_t has to satisfy additional conditions as in Chap. 7 of [6]. We then get a time scaled version of the same limiting O.D.E. in the asynchronous case as the synchronous case and the asymptotic behavior of the algorithm is the same as that for the synchronous case.

3.2   Randomized Kaczmarz Approach

Another asynchronously distributed approach is to apply the randomized Kacz-marz algorithm (RK-approach) to (2). We solve the linear system (2) of the form $\underline{A}x = \underline{b}$, where $\underline{A} = (\mathcal{A} + \mathcal{A}^T) + \mu(\mathcal{B} + \mathcal{B}^T)$ and $\underline{b} = \mu(\mathcal{B} + \mathcal{B}^T)Y$. Let $a_i :=$ the ith row of $\underline{A}$, $\hat{a}_i :=$ the unit normalized row a_i and $p_i :=$ the probability of sampling it for the Kaczmarz update (e.g., $p_i = \frac{1}{N}$ for uniform sampling). The update rule at step t is given by

$$x(t+1) = x(t) + \sum_i I\{\xi(t) = i\} \frac{b(i) - \langle a_i, x(t)\rangle}{\|a_i\|_2^2} a_i^T, \tag{5}$$

where $Prob(\xi(t) = i) = p_i \; \forall \; i \in \{1, 2, ..., N\}$. Let λ_{min} be the smallest non-negative eigenvalue of the matrix $\sum_i p_i \hat{a}_i^T \hat{a}_i$. From [7], we have that for $\lambda_{min} \in (0, 1)$, $x(t) - x^* \to 0$ almost surely and $E[\|x(t) - x^*\|^2] \to 0$ exponentially with rate $(1 - \lambda_{min})$ where x^* is such that $\underline{A}x^* = \underline{b}$.

The randomized Kaczmarz scheme is an inherently distributed and asyn-chronous scheme because only one component (corresponding to a single node) is being updated at a time, using local information from the node's neighbours. This sampling is done probabilistically in an i.i.d. fashion with probability vector $p = [p_1, ..., p_N]^T$. We may drop the assumption of identical distribution sampling and still retain exponential decay of mean square error as long as λ_{min} remains bounded away from 1 from above for time-varying p. In fact, one may drop the independence assumption as well, replacing it with the above condition on the conditional sampling probabilities given the history. The exponential decay rate of mean square error then is $1 - \lambda^*$, where $\lambda^* < 1$ is the aforementioned upper bound. For normal matrices, the condition number of a matrix M is given by $\kappa(M) = \left|\frac{\lambda_{max}(M)}{\lambda_{min}(M)}\right|$, where $\lambda_{min}(M)$ and $\lambda_{max}(M)$ are its minimum and max-imum eigenvalues, respectively. Hence a high condition number would imply a very small λ_{min} resulting in very slow convergence.

3.3   Comments on Implementation

For the asynchronous implementation of (4) and (5) we have a Poisson local clock at each node. The node updates its classification function once the local clock ticks. We simulate this by performing a coin toss at each instant of the global clock. The coin has a low probability p_H of turning up heads. The node updates its classification function if it gets head at that time instant. There can be multiple nodes updating at a given global instant while there can also be no nodes updating at some other instant. This leads to the nodes updating their classification function asynchronously.

We emphasize that both our approaches, SA and RK, require the information only from the neighbors of the node. In fact, for any update, SA needs the information only from one random neighbor. Updating each node independent of the other nodes with only local information implies the distributed nature of the updates. In theory, we can have different p_H for each node and convergence is assured if p_H for all nodes is bounded away from zero. What we find particularly interesting is that the unlabelled nodes do not need to know the location of the labelled nodes. The information from the labelled nodes propagates through the network from neighbor to neighbor.

4 Application to Specific SSL Methods

We discuss the application of the two approaches to several well known SSL methods.

4.1 Normalized Laplacian-Type Methods

For $\mathcal{A} = D^{\sigma-1}LD^{\sigma-1}$ and $\mathcal{B} = D^{2\sigma-1}$ in (2), we obtain the normalized Laplacian-type methods [1,2,25,26] which yields the following equation on simplification:

$$\left(I - \frac{1}{1+\mu}D^{-\sigma}AD^{\sigma-1}\right)F_{*k} = \frac{\mu}{1+\mu}Y_{*k}. \tag{6}$$

This equation can be solved using RK-approach or can be recast to resemble (3) as follows:

$$F_{*k} = \frac{1}{1+\mu}D^{-\sigma}AD^{\sigma-1}F_{*k} + \frac{\mu}{1+\mu}Y_{*k}. \tag{7}$$

In case of normalized Laplacian-type methods, $\widetilde{B} = \frac{1}{1+\mu}D^{-\sigma}AD^{\sigma-1}$ is similar to and a scaled version of the transition probability matrix $D^{-1}A$. Hence its top eigenvalue is less than 1. Thus a stochastic approximation of the form of (4) written for (7) will converge to its solution.

4.2 Regularized Laplacian Method

$\mathcal{A} = L, \mathcal{B} = I$ in (2) yield the regularized Laplacian method [3,9,21] which on simplification gives:

$$(L + \mu I)F_{*k} = \mu Y_{*k}. \tag{8}$$

This can be solved using RK-approach or can be rewritten in a form similar to Eq. (3) which leads to the stochastic approximation scheme:

$$F_{*k} = (D + \mu I)^{-1}AF_{*k} + \mu(D + \mu I)^{-1}Y_{*k}. \tag{9}$$

In case of regularized Laplacian, $\widetilde{B} = (D+\mu I)^{-1}A$. Its row sums are <1, making $\widetilde{B}$ sub-stochastic with its top eigenvalue <1. Hence a stochastic approximation for (9) of the form (4) will converge to its solution.

4.3 Harmonic Functions Method

Finally, substituting in (2) $\mathcal{A} = L$ and $\mathcal{B} = diag(I,0)$, where the non-zero elements correspond to the labelled points, we get the harmonic functions method [27] which on simplification gives:

$$(L + \mu\, diag(I,0))F_{*k} = \mu Y_{*k}. \tag{10}$$

A direct application of SA-approach may not be possible for this, but application of the RK-approach is straightforward.

5   Experiments

We test our distributed approaches on the three methods (normalized Laplacian with $\sigma = 1/2$ [25], regularized Laplacian [3,9,21] and harmonic functions [27]) applied to synthetic as well as real world graphs. As performance metric, we use classification error, i.e., the percentage of nodes wrongly classified. For all the experiments we took $\mu = 0.5$ as the regularization parameter (the results are robust for reasonable values of μ). For comparison purposes, we take 1 iteration $=$ N steps, where N is the number of nodes in the graph and one step is defined as one application of formula (5) for RK-approach and formula (4) for SA-approach, respectively. In most experiments, a uniform distribution is used for sampling rows in RK-approach. A decreasing step size for node i of $\frac{1}{2+\nu(t,i)}$ where $\nu(t,i)$ is the local clock at node i is used in (4).

5.1   WebKB Graph

We look at the classification of webpages of 4 universities - Cornell, Texas, Washington and Wisconsin - corresponding to the popular WebKB dataset [10]. The graph formed by the hyperlinks connecting these pages is taken such that only webpages with hyperlinks to webpages within the dataset are considered. Self-links are removed. Clusters thus formed are: Cornell (676), Texas (590), Washington (982) and Wisconsin (613). The highest degree node from each class (university main web page) is labelled. Figure 1a shows the error evolution for the Kaczmarz implementation of the three SSL methods. Convergence only occurs for normalized Laplacian for the number of iterations shown while both other methods have a negative slope. Convergence occurs for regularized Laplacian method after 200 iterations. All the methods seem to have the same initial rate of convergence however they change drastically after around 3 iterations. Theoretical Classification Error (TCE) for all the methods is almost the

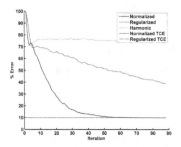
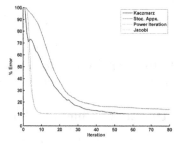

(a) Comparison of Kaczmarz implementation of all the three methods. In dotted is shown the theoretical classification error (TCE).

(b) Comparison of Kaczmarz, power iteration, Jacobi and stochastic approximation implementation of normalized Laplacian.

Fig. 1. Performance on the WebKB graph

same. Figure 1b shows the comparison of RK-approach, SA-approach, distributed Jacobi and power iteration for normalized Laplacian. Power iteration is simply the repeated application of (7) with one iteration defined as multiplication with $\frac{1}{1+\mu}D^{-1/2}AD^{-1/2}$ once. RK-approach's performance is better than that of SA-approach for this graph in terms of error while convergence occurs faster for the latter. Convergence is fastest for power iterations (PI) implementation since in each iteration, the classification function of all the nodes is being updated and as a result, the update performed in the consequent iteration would be with the updated classification functions of the neighbors. This is unlike SA and RK approach where the update might use the non-updated classification function of its neighbor.

Let a call be defined as the transfer of information to a node from its neighbor. This information is used for updating the classification function. As Fig. 1b shows, the number of iterations is almost the same for RK and SA. While updating the classification function of a node in one step of SA, only one of its neighbors is called, whereas all neighbors are called in RK and Jacobi. As a result, RK and Jacobi have more exchange of information and is computationally more expensive than SA.

5.2 US Football Graph

We next see the classification in US college football network [12]. Nodes in the graph represent colleges that participated in the Division 1 games for the 2000 season. Edges between nodes represent games between the two teams they connect. The classes and the nodes corresponding to each class are known for this graph. There are 12 classes with each class consisting of 8–12 teams. Teams within the same class, called 'conference', tend to play more games with each other than with teams from another conference. The highest degree node from each class (conference) is labelled. Figure 2a shows the error evolution for the

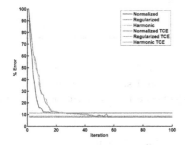
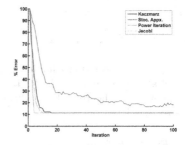

(a) Comparison of Kaczmarz implementation of all the three methods. In dotted is shown the theoretical classification error (TCE).

(b) Comparison of Kaczmarz, power iteration, Jacobi and stochastic approximation implementation of normalized Laplacian.

Fig. 2. Performance on the US Football graph

RK-approach implementation of the three methods. The convergence is fastest for the normalized Laplacian while the error is less for regularized Laplacian and harmonic functions methods. Figure 2b shows the comparison of RK-approach, SA-approach, distributed Jacobi and power iteration for normalized Laplacian. In this example, RK and Jacobi show similar behaviour. We recall again that RK and Jacobi use more information and computationally more expensive than SA.

5.3 Gaussian Mixture Model Graph

A Gaussian mixture model graph of 10000 nodes with 3 classes was created with the probability of a node belonging to either of the three classes being equal. Each class was generated on a Gaussian kernel. Nodes within a given radius of each other shared edges. Two nodes with the highest degree from each class were labelled. Figure 3a compares the error evolution of RK-approach for the three methods. Convergence only occurs for normalized Laplacian method for the number of iterations shown, while the other methods have a negative slope. All the methods seem to have the same initial rate of convergence which changes drastically after a few iterations. Figure 3b compares the RK-approach, SA-approach and power iteration for normalized Laplacian. The performance of PI is best in terms of both error and rate of convergence. We would like to emphasize that we sacrifice the rate of convergence for the distributed nature of algorithms. However, the performance for RK and SA approaches for this graph is different as compared to WebKB graph, Fig. 1b. RK-approach has a higher error as well as convergence time as compared to SA-approach.

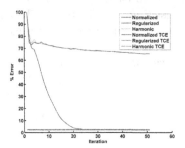

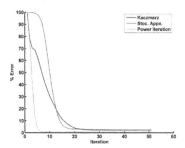

(a) Comparison of Kaczmarz implementation of all the three methods. In dotted is shown the theoretical classification error (TCE).

(b) Comparison of Kaczmarz, power iteration and stochastic approximation implementation of normalized Laplacian.

Fig. 3. Performance on Gaussian mixture model graph of 10000 nodes.

5.4 Online Learning

In the RK-approach as well as SA-approach, the classification function is updated only for one or few nodes in one step. In other words, only local information is used each time while updating, allowing for natural application of our approaches to dynamic setting with streaming data. To illustrate the performance of our approaches in the dynamic setting, we consider a dynamic stochastic block model graph in which nodes enter and leave the graph. Upon arrival, a node is connected with another node of the same class with probability $p_{in} = 0.15$ and node from a different class with probability $p_{out} = 0.01$; $p_{in} \gg p_{out}$. Nodes arrive into the graph according to the Poisson process with rate λ_{arr}. The class of the arriving node is chosen according to a pre-specified probability distribution. Each node stays in the graph for a random time that is exponentially distributed with mean $1/\mu_{dep}$, after which it leaves. The maximum number of nodes in the graph is limited to K, i.e., nodes arriving when the number of nodes in the graph is K do not become a part of the graph. This system can be modelled as an M/M/K/K queue. As a result, irrespective of the number of nodes that the graph has initially, the average of the number of nodes in the graph will reach a steady state value given approximately by $\frac{\lambda_{arr}}{\mu_{dep}}$. In the considered example, the graph has 3 classes and an incoming node could belong to either of the classes with equal probability. We choose $K = 1000$, $\lambda_{arr} = 1/(2 \times 10^4)$ and $\mu_{dep} = 1/10^7$. Therefore, $\frac{\lambda_{arr}}{\mu_{dep}} = 500$. Two nodes with the maximum degree from each class were chosen as the labelled nodes during initialization. In case a labelled node left, then a random neighbor was labelled. In the plots, K steps are considered as one iteration.

Figures 4a, b and c show the error evolution of the RK-approach implementation of normalized Laplacian, regularized Laplacian and harmonic functions methods, respectively, for the dynamic stochastic block model graph. The variation of the graph size is also shown in the same figures. In terms of accuracy, the performances of regularized Laplacian and harmonic function methods are similar, being between 0–1.5 % during steady state. Interestingly, normalized Laplacian method has a higher error compared to the other two, being close to 3.5 %. Figure 5 shows a zoom of the error evolution from Figs. 4a, b and c. From this figure, it can be seen that the convergence is faster for the normalized Laplacian compared to the other two methods, both of which have almost the same convergence time.

(a) Normalized Laplacian. (b) Regularized Laplacian. (c) Harmonic functions.

Fig. 4. Error evolution and graph size variation for a dynamic stochastic block model graph for Kaczmarz implementation of various methods.

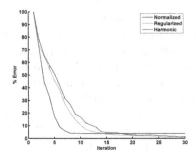

Fig. 5. This plot shows the comparison of Kaczmarz implementation of different methods for a dynamic stochastic block model.

5.5   Faster Convergence for Normalized Laplacian

The convergence was faster for the normalized Laplacian method compared to the regularized Laplacian and harmonic functions methods in all the cases of the RK-approach application. This can be understood from the condition number values for each method on different graphs as shown in Table 1 and invoking theoretical observations from Sect. 3.2. The condition number is the smallest for the normalized Laplacian method and the highest for the harmonic functions method in all the cases. λ_{max} being less than 1 for all the methods, the λ_{min} must be very small for the harmonic functions method as compared to normalized Laplacian to explain the large condition number. This leads to their large convergence times.

Table 1. Condition number values.

	Normalized	Regularized	Harmonic
US Football	3.88	32.32	314.55
WebKB	5	300	5.6×10^{18}
Gaussian	4.46	2.8×10^3	4.6×10^6

6   Conclusion

We proposed two asynchronously distributed approaches for graph-based semi-supervised learning. The first approach is based on stochastic approximation, whereas the second is based on the randomized Kaczmarz algorithm. We demonstrated that both the approaches can be naturally applied online to streaming data. Both were analysed theoretically and by experiments. Our main conclusions: there is no clear winner in terms of accuracy but the SA-approach generally outperformed the RK-approach in terms of operations count. In terms of accuracy, RK-approach performed better on real world datasets (US football and

WebKB) while SA-approach performed better on the synthetic Gaussian mixture model. When using RK-approach, normalized Laplacian method showed much faster convergence as compared to regularized Laplacian and harmonic functions owing to its low condition number.

Acknowledgement. This work was supported by CEFIPRA grant no. 5100-IT1 "Monte Carlo and Learning Schemes for Network Analytics," and Inria Nokia Bell Labs.

References

1. Avrachenkov, K., Dobrynin, V., Nemirovsky, D., Pham, S.K. Smirnova, E.: PageRank based clustering of hypertext document collections. In: Proceedings of ACM SIGIR (2008)
2. Avrachenkov, K., Gonçalves, P., Mishenin, A., and Sokol, M.: Generalized optimization framework for graph-based semi-supervised learning. In: Proceedings of SDM (2012)
3. Avrachenkov, K., Chebotarev, P., Mishenin, A.: Semi-supervised learning with regularized Laplacian. Accepted in Optimization Methods & Software (2016)
4. Bengio, Y., Delalleau, O., Le Roux, N.: Label propagation and quadratic criterion. In: Semi-supervised Learning, ch. 10 (2006)
5. Bertsekas, D.P., Tsitsiklis, J.N.: Parallel and Distributed Computation: Numerical Methods. Prentice Hall, Englewood Cliffs (1989)
6. Borkar, V.S.: Stochastic Approximation: A Dynamical Systems Viewpoint. Hindustan Publishing Agency, Cambridge University Press, New Delhi, Cambridge (2008)
7. Borkar, V.S., Karamchandani, N., Mirani, S.: Randomized Kaczmarz for rank aggregation from pairwise comparisons. In: IEEE ITW (2016)
8. Chapelle, O., Schölkopf, B., Zien, A.: Semi-supervised Learning. MIT Press, London (2006)
9. Chebotarev, P., Shamis, E.: The matrix-forest theorem and measuring relations in small social groups. Autom. Remote Control **58**(9), 1505–1514 (1997)
10. Craven, M., McCallum, A., PiPasquo, D., Mitchell, T., Freitag, D.: Learning to extract symbolic knowledge from the World Wide Web (No. CMU-CS-98-122). School of computer Science, Carnegie-Mellon University, Pittsburgh, PA (1998)
11. Fouss, F., Francoisse, K., Yen, L., Pirotte, A., Saerens, M.: An experimental investigation of kernels on graphs for collaborative recommendation and semisupervised classification. Neural Netw. **31**, 53–72 (2012)
12. Girvan, M., Newman, M.E.J.: Community structure in social and biological networks. PNAS USA **99**, 7821–7826 (2002)
13. Gleich, D.F., Mahoney, M.W.: Using local spectral methods to robustify graph-based learning algorithms. In: Proceedings of ACM SIGKDD (2015)
14. Gower, R.M., Richtárik, P.: Randomized iterative methods for linear systems. SIAM J. Matrix Anal. Appl. **36**(4), 1660–1690 (2015)
15. Ito, T., Shimbo, M., Kudo, T., Matsumoto, Y.: Application of kernels to link analysis. In: Proceedings of ACM SIGKDD (2005)
16. Liu, J., Wright, S.J., Sridhar, S.: An asynchronous parallel randomized Kaczmarz algorithm (2014). arXiv preprint: arXiv:1401.4780

17. Needell, D., Ward, R., Srebro, N.: Stochastic gradient descent, weighted sampling, and the randomized kaczmarz algorithm. In: Proceedings of NIPS (2014)
18. Pan, J.J., Pan, S.J., Yin, J., Ni, L.M., Yang, Q.: Tracking mobile users in wireless networks via semi-supervised colocalization. IEEE Trans. Pattern Anal. Mach. Intell. **34**(3), 587–600 (2012)
19. Ravi, S., Diao, Q.: Large scale distributed semi-supervised learning using streaming approximation. In: Proceedings of AISTATS (2016)
20. Shivanna, R., Chatterjee, B.K., Sankaran, R., Bhattacharyya, C., Bach, F.: Spectral norm regularization of orthonormal representations for graph transduction. In: Advances in Neural Information Processing Systems, pp. 2215–2223 (2015)
21. Smola, A.J., Kondor, R.: Kernels and regularization on graphs. In: Schölkopf, B., Warmuth, M.K. (eds.) COLT-Kernel 2003. LNCS (LNAI), vol. 2777, pp. 144–158. Springer, Heidelberg (2003). doi:10.1007/978-3-540-45167-9_12
22. Strohmer, T., Vershynin, R.: A randomized Kaczmarz algorithm with exponential convergence. J. Fourier Anal. Appl. **15**(2), 262–278 (2009)
23. Talukdar, P.P., Crammer, K.: New regularized algorithms for transductive learning. In: Buntine, W., Grobelnik, M., Mladenić, D., Shawe-Taylor, J. (eds.) ECML PKDD 2009. LNCS (LNAI), vol. 5782, pp. 442–457. Springer, Heidelberg (2009). doi:10.1007/978-3-642-04174-7_29
24. Valko, M., Kveton, B., Huang, L., Ting, D.: Online semi-supervised learning on quantized graphs. In: Proceedings of UAI (2010)
25. Zhou, D., Bousquet, O., Lal, T.N., Weston, J., Schölkopf, B.: Learning with local and global consistency. Adv. Neural Inf. Process. Syst. **16**, 321–328 (2004)
26. Zhou, D., Burges, C.J.: Spectral clustering and transductive learning with multiple views. In: Proceedings of ICML (2007)
27. Zhu, X., Ghahramani, Z., Lafferty, J.: Semi-supervised learning using Gaussian fields and harmonic functions. In: Proceedings of ICML (2003)
28. Zhu, X.: Semi-supervised learning: literature survey. University of Wisconsin-Madison Research report TR 1530 (2005)
29. Zhu, X., Goldberg, A.B.: Introduction to Semi-supervised Learning. Morgan & Claypool, San Rafael (2009)
30. Zouzias, A., Freris, N.M.: Randomized gossip algorithms for solving Laplacian systems. In: Proceedings of ECC (2015)

Existence and Region of Critical Probabilities in Bootstrap Percolation on Inhomogeneous Periodic Trees

Milan Bradonjić[1(✉)] and Stephan Wagner[2]

[1] Mathematics of Systems, Nokia Bell Labs, Murray Hill, NJ 07974, USA
`milan@research.bell-labs.com`
[2] Department of Mathematical Sciences, Stellenbosch University,
Private Bag X1 Matieland, Stellenbosch 7602, South Africa
`swagner@sun.ac.za`

Abstract. Bootstrap percolation is a growth model inspired by cellular automata. At the initial time $t = 0$, the bootstrap percolation process starts from an initial random configuration of active vertices on a given graph, and proceeds deterministically so that a node becomes active at time $t = 1, 2, \ldots$ if sufficiently many of its neighbors are already active at the previous time $t - 1$. In the most basic model, all vertices have the same initial probability of being active in the initial configuration. One of the main questions is to determine the percolation threshold (if it exists) with the property that all nodes in the given graph become active asymptotically almost surely (a.a.s.) for the initial probability above this threshold, while this is not the case below the threshold. In this work, we study a scenario where the nodes do not all receive the same probabilities, but to keep the problem tractable, we impose conditions on the shape of the graph and the initial probabilities. Specifically, we consider infinite periodic trees, in which the degrees and initial probabilities of nodes on a path from the root node are periodic, with a given periodicity. Instead of the simple percolation threshold, we now obtain an entire region of possible probabilities for which all nodes in the tree become a.a.s. active. We show: (i) that the unit cube, as the support of the initial probabilities, can be partitioned into two regions, denoted by W_0 and $\overline{W}_0$, such that the tree becomes (does not become) a.a.s. fully active for any initial probability vector that belongs to $\overline{W}_0$ (resp. W_0); (ii) for every node in the tree, we provide the probability that the node becomes eventually active, for any initial probability vector that belongs to W_0; (iii) further, we specify the boundary of W_0 and show how it can be numerically computed.

1 Introduction

In classical percolation theory, nodes of a graph become active according to certain probabilities to form a static configuration. *Bootstrap percolation* is a

S. Wagner—Supported by the National Research Foundation of South Africa under grant number 96236.

© Springer International Publishing AG 2016
A. Bonato et al. (Eds.): WAW 2016, LNCS 10088, pp. 47–59, 2016.
DOI: 10.1007/978-3-319-49787-7_5

variant inspired by cellular automata that proceeds dynamically afterwards: starting from an initial configuration (determined in the same way as in classical percolation), the process proceeds in discrete time-steps, where a node is active at time t if it is or sufficiently many of its neighbors are already active at time $t - 1$. It may thus happen that all nodes become eventually active, which poses the natural question for the existence of a critical threshold probability (assuming that all nodes have the same initial probability of being active) such that all nodes (do not) become a.a.s. active if the initial probability is greater (resp. smaller) than the threshold. There is a large body of work on bootstrap percolation on different graph models: regular or irregular, discrete or random, homogeneous or inhomogeneous, as well as in isotropic or anisotropic environment [1–13, 15–25].

In this work, we study a scenario where the nodes do not all receive the same probabilities, but to keep the problem tractable, we impose conditions on the shape of the graph and the probabilities and activation thresholds. Specifically, we consider infinite periodic trees, in which the degrees of nodes on a path from the root node are periodic, and also impose a periodicity condition on the probabilities. The existence of a percolation threshold for *periodic trees* has been established in [14] in the case where the initial probability of being active and the activation threshold are the same for all nodes.

In this work the initial probability and the activation threshold are functions of the node itself, which is the main generalization of the model analyzed in [14]. As motivation, consider a dynamical process (e.g. advertisement, rumor, or viral spread). It is usually the case that particles in the system become initially active (e.g. obtain the initial piece of information or become infected) with different probabilities, as well as that the activation threshold differs among particles (e.g. the level required to convince a customer to buy a new product or for one to become infected depends on the individual itself).

Hence in this work we consider the following object. A periodic tree corresponding to a sequence $d_0, d_1, \ldots, d_{\ell-1}$ is an infinite tree with a root node such that every vertex at distance i mod ℓ from the root has degree $d_i + 1$. In addition to the degrees, we specify activation thresholds $\theta_0, \theta_1, \ldots, \theta_{\ell-1}$ ($2 \leqslant \theta_i \leqslant d_i - 1$) for the bootstrap percolation. This means that a node at distance i mod ℓ from the root will become active: either at the initial random phase $t = 0$, or once θ_i of its neighbors are active at the previous time step. Finally, and this is the main difference to all prior work, we allow the initial probabilities to be periodic as well (rather than fixed throughout the tree): at time 0, a node at distance i mod ℓ from the root becomes active with probability p_i. Note that the periods of d_i, θ_i and p_i do not a priori have to be equal, but we can assume so without loss of generality, since we can otherwise replace ℓ by the least common multiple of the periods.

In this work, we study bootstrap percolation on inhomogeneous periodic trees with different initial probabilities. Instead of the simple percolation threshold, as shown in [14], we now obtain an entire region of possible probabilities for which all nodes in the tree become a.a.s. active. We show: (i) that the unit cube, as the support of the initial probabilities $p_0, p_1, \ldots, p_{\ell-1}$, can be partitioned into two regions, denoted by W_0 and $\overline{W}_0$, such that: the tree becomes (does not become) a.a.s. fully active for any initial probability vector that belongs to $\overline{W}_0$ (resp. W_0); (ii) for every node in the tree, we provide the probability that the node becomes eventually active, for any initial probability vector that belongs to W_0; (iii) further, we specify the boundary of W_0 and how it can be numerically computed. In fact we derive the explicit system of equations from which one can numerically compute the boundary of W_0.

2   Definitions and Preliminaries

Formally, bootstrap percolation is a cellular automaton defined on an underlying graph G with state space $\{0,1\}^{V(G)}$ whose initial configuration is chosen by a Bernoulli product measure. In other words, every node is in one of two different states 0 or 1, *inactive* or *active* respectively, and a node v is active with some initial probability p_v, independently of other nodes, within the initial configuration at $t = 0$. In this work the initial probability p_v is a function of the node itself. After drawing an initial configuration, a discrete time deterministic process updates the configuration according to a local rule: an inactive node v becomes active at time $t + 1$ if the number of its active neighbors at t is greater than or equal to some specified *activation parameter* θ_v, which is a function of the node v as well. Once an inactive node becomes active it remains active. A configuration that does not change at the next time step is a *stable* configuration. A configuration is *fully active* if all its nodes are active.

In this work we study the bootstrap percolation process on periodic trees defined as follows.

Definition 1 (Periodic tree). *Let $\ell, d_0, d_1, \ldots, d_{\ell-1} \in \mathbb{N}$. An ℓ-periodic tree $\mathbb{T}_{d_0, d_1, \ldots, d_{\ell-1}}$ is defined as follows. Consider a node v_0, called root. The nodes at distance i mod ℓ from v_0 have degree $d_i + 1$ for $i \in \mathbb{N}_0$. In particular, the root has degree $d_0 + 1$.*

An infinite d-regular tree is a special case: a 1-periodic tree where each node has degree $d + 1$.

We also need to define the following oriented tree.

Definition 2 (Oriented periodic tree). *Let $\ell, d_0, d_1, \ldots, d_{\ell-1} \in \mathbb{N}$. An oriented ℓ-periodic tree $\vec{\mathbb{T}}_{d_0, d_1, \ldots, d_{\ell-1}}$ is defined as follows. Consider a node v_0, called root. The nodes at distance i mod ℓ from v_0 have in-degree d_i and out-degree 1 for $i \in \mathbb{N}_0$, except for the root, which has out-degree 0.*

We note that an oriented ℓ-periodic tree is a periodic tree with all edges oriented towards the root, the exception being the root degree.

Definition 3. *For $\vec{p} = (p_0, p_1, \ldots, p_{\ell-1})$ define*

$$\vec{p}_{-i} := (p_0, p_1, \ldots, p_{i-1}, p_{i+1} \ldots, p_{\ell-1}) \tag{1}$$

to be a vector obtained by erasing the i-th coordinate from $\vec{p}$. Conversely, $\vec{p}$ is obtained by inserting p_i at the i-th coordinate in $\vec{p}_{-i}$, which we write as

$$\vec{p} \equiv (\vec{p}_{-i} | p_i). \tag{2}$$

2.1   Notation

Throughout this work, given $d_0, d_1, \ldots, d_{\ell-1}$, we will usually use $\mathbb{T}$ (resp. $\vec{\mathbb{T}}$) as a shorthand for $\mathbb{T}_{d_0, d_1, \ldots, d_{\ell-1}}$ (resp. $\vec{\mathbb{T}}_{d_0, d_1, \ldots, d_{\ell-1}}$).

Given a tree $\mathbb{T} = \mathbb{T}_{d_0, d_1, \ldots, d_{\ell-1}}$, partition the node set $V(\mathbb{T})$ into ℓ *classes* of nodes V_i, where V_i contains the nodes of $\mathbb{T}$ at distance $i \bmod \ell$ from the root with degree $d_i + 1$, activation threshold θ_i, and initial probability p_i, for $i \in \{0, 1, \ldots, \ell - 1\}$.

In the following, all indices will be considered modulo ℓ, e.g. $x_\ell = x_0$.

3   Bootstrap Percolation

This section is devoted to bootstrap percolation on periodic inhomogeneous trees. The main result is given by Theorems 1 and 2, showing the regions of initial probabilities for which the tree $\vec{\mathbb{T}}$, respectively $\mathbb{T}$, become a.a.s. fully active. Finally, we show that these two regions are identical.

Functions of the form

$$\varphi_{d,p,\theta}(x) := p + (1-p) \sum_{k=\theta}^{d} \binom{d}{k} x^k (1-x)^{d-k}, \tag{3}$$

where $p \in [0, 1]$, will play a key role, as they capture one time step in bootstrap percolation. Intuitively, the first term stands for the probability of a node to be initially active, the sum in the second term for the probability of becoming active because at least θ of its neighbors are. The following result appeared in different forms in [11,17] and the proof is given in [14, Lemma 2.1].

Lemma 1. *Given $d, \theta \in \mathbb{N}$ such that $2 \leqslant \theta \leqslant d - 1$ and $p \in [0, 1]$, there exists $p_c \in (0, 1)$ such that for any $p > p_c$ we have $\varphi_{d,p,\theta}(x) > x$ for every $x \in (0, 1)$, and 1 is the only solution of $\varphi_{d,p,\theta}(x) = x$ in $[0, 1]$. For $p < p_c$, there are two solutions in $[0, 1]$ other than 1, and for $p = p_c$ there is one other solution (of multiplicity 2).*

3.1   Bootstrap Percolation on an Oriented Tree $\vec{\mathbb{T}}$

Following the methodology of [17], we first show the existence of a threshold region for oriented trees, i.e. a region W_0 of probabilities p_i for which not all nodes become active asymptotically almost surely (a.a.s), while for a choice of probabilities in the complement $\overline{W}_0$, all nodes become active a.a.s. Later we show that the regions for oriented and unoriented periodic trees with the same parameters are actually the same.

The dynamics must be defined in a slightly different way for bootstrap percolation, though (in other words, we need to define an *oriented* version of bootstrap percolation): a node in class i becomes newly active if θ_i of its in-neighbors (neighbors for which the orientation of the associated edge is towards the node) are active in the previous step. For this slightly modified version, we obtain the following result:

Theorem 1. *Consider bootstrap percolation on an oriented tree $\vec{\mathbb{T}}$ with parameters $(d_i, \theta_i, p_i)_{i=0}^{\ell-1}$, where $2 \leqslant \theta_i \leqslant d_i - 1$. Let W_0 be the set of probability vectors $\vec{p} \in [0,1]^\ell$ such that there exists a solution in $[0,1]^\ell$ of the system*

$$x_i = p_i + (1 - p_i) \sum_{k=\theta_i}^{d_i} \binom{d_i}{k} x_{i+1}^k (1 - x_{i+1})^{d_i - k} \tag{4}$$

that is strictly less than $\vec{1}$, i.e. $x_i < 1$ for every $i \in \{0, 1, \ldots, \ell - 1\}$. Then (i) for every $\vec{p} \in W_0$, $\vec{\mathbb{T}}$ does not become fully active a.a.s.; (ii) for every $\vec{p} \in \overline{W}_0 := [0,1]^\ell \setminus W_0$, $\vec{\mathbb{T}}$ becomes fully active a.a.s. Moreover, there exist constants $\delta, \sigma \in (0,1)$ such that $W_0 \supset [0,\delta]^\ell$ and $W_0 \subset [0,\sigma]^\ell$.

The dynamics of the bootstrap percolation process on $\vec{\mathbb{T}}$ are captured by knowing the states of every node $v \in V_i$, in every class V_i, at every time $t \in \mathbb{N}_0$. These states are denoted by $\vec{\zeta}_{i,t}(v) \in \{0,1\}$.

It is intuitive that the higher p_i, the higher the probability that a node in class i becomes eventually active. Also, if all p_i are equal to 0, the system is already in a state of equilibrium, where the state of every node in the tree is 0 (inactive). On the other hand, if all p_i are equal to 1, the system is in yet another equilibrium, where the state of every node in the tree is 1 (active).

3.2 Proof of Theorem 1

The initial steps in the proof are analogous to those in [14, 17]. However, the main difference is that we consider different degrees d_i, different activation thresholds θ_i and most importantly different initial probabilities p_i.

For every class V_i, choose any node $v \in V_i$. Conditioning upon whether this node v was active at time 0 or not (i.e., $\vec{\zeta}_{i,0}(v) = 0$ or $\vec{\zeta}_{i,0}(v) = 1$), the probability that the node v is active at time t is given by

$$\mathbb{P}\left(\vec{\zeta}_{i,t}(v) = 1\right) = \mathbb{P}\left(\vec{\zeta}_{i,0}(v) = 1\right) + \mathbb{P}\left(\vec{\zeta}_{i,0}(v) = 0\right) \mathbb{P}\left(\sum_{u \leadsto v} \vec{\zeta}_{i+1,t-1}(u) \geqslant \theta_i\right),$$

where the symbol "$\leadsto$" indicates that u is a neighbor of v in the oriented tree $\vec{\mathbb{T}}$ and the edge orientation is from u to v.

Given symmetry and the dynamical rules of the bootstrap percolation process, the $\vec{\zeta}_{i+1,t-1}(u)$ in the equation above are independent Bernoulli random variables with the same distribution; moreover, they are independent of $\vec{\zeta}_{i,0}(v)$. Introducing $\vec{z}_{i,t} := \mathbb{P}\left(\vec{\zeta}_{i,t}(v) = 1\right)$, we obtain the following system of recurrence equations:

$$\vec{z}_{i,t} = p_i + (1 - p_i) \sum_{k=\theta_i}^{d_i} \binom{d_i}{k} \vec{z}_{i+1,t-1}^k (1 - \vec{z}_{i+1,t-1})^{d_i - k}, \tag{5}$$

for $i = 0, 1, \ldots, \ell - 1$.

In order to simplify the notation, for given parameters $(d_i, \theta_i, p_i)_{i=0}^{\ell-1}$ that characterize a tree $\vec{\mathbb{T}}$ (as well as $\mathbb{T}$), we define the auxiliary functions $\phi_i(x; p_i) := \varphi_{d_i, p_i, \theta_i}(x)$ on $[0, 1]$, i.e.

$$\phi_i(x; p_i) := p_i + (1 - p_i) \sum_{k=\theta_i}^{d_i} \binom{d_i}{k} x^k (1 - x)^{d_i - k}, \tag{6}$$

and the binomial tail

$$B_i(x) := \sum_{k=\theta_i}^{d_i} \binom{d_i}{k} x^k (1 - x)^{d_i - k}. \tag{7}$$

Now, the recurrence system (5) can be rewritten as

$$\vec{z}_{i,t} = \phi_i(\vec{z}_{i+1,t-1}; p_i), \tag{8}$$

for $i = 0, 1, \ldots, \ell - 1$, and all $\vec{z}_{i,t}$ belong to $[0, 1]$ since ϕ_i maps $[0, 1]$ to $[0, 1]$.

Claim. For every i, $\vec{z}_{i,t}$ is non-decreasing in t.

Proof. For $t = 0$, $\vec{z}_{i,0} = p_i$. From (5) $\vec{z}_{i,1} \geqslant p_i$, thus the claim holds for $t = 0$. Assume that for some t and every i, $\vec{z}_{i,t} \geqslant \vec{z}_{i,t-1}$. $B_i(x)$ is increasing in x. Thus

$$\vec{z}_{i,t+1} = \phi_i(\vec{z}_{i+1,t}; p_i) = p_i + (1 - p_i) B_i(\vec{z}_{i+1,t}) \geqslant p_i + (1 - p_i) B_i(\vec{z}_{i+1,t-1}) = \vec{z}_{i,t},$$

and the statement follows by mathematical induction.

So the $\vec{z}_{i,t}$ are non-decreasing in t and belong to $[0, 1]$, thus by the monotone convergence theorem the limits $\vec{z}_{i,\infty} := \lim_{t \to \infty} \vec{z}_{i,t}$ exist, and they lie in $[0, 1]$. By (8),

$$\vec{z}_{i,\infty} = \phi_i(\vec{z}_{i+1,\infty}; p_i), \tag{9}$$

for all $i \in \{0, 1, \ldots \ell - 1\}$. At this moment, we introduce the vector of the limiting values for $t \to \infty$:

$$\vec{z}_\infty := (\vec{z}_{0,\infty}, \vec{z}_{1,\infty}, \ldots, \vec{z}_{\ell-1,\infty}) \tag{10}$$

as well as the original ones at time $t = 0$:

$$\vec{p} := (p_0, p_1, \ldots, p_{\ell-1}) = (\vec{z}_{0,0}, \vec{z}_{1,0}, \ldots, \vec{z}_{\ell-1,0}). \tag{11}$$

Applying (9) ℓ times, for every i, we obtain the equations of one variable

$$\vec{z}_{i,\infty} = \phi_i(\phi_{i+1}(\cdots(\phi_{i-1}(\vec{z}_{i,\infty}; p_{i-1})\cdots); p_{i+1}); p_i) = F_i(\vec{z}_{i,\infty}), \tag{12}$$

where we define

$$F_i := \phi_i \circ \phi_{i+1} \circ \cdots \circ \phi_{\ell-1} \circ \phi_0 \circ \cdots \circ \phi_{i-1}. \tag{13}$$

Notice that by (5) and (9), $\vec{z}_\infty \neq \vec{0}$ if and only if $\vec{p} \neq 0$. Next, we show that there exists a non-empty hypercube $[0, \delta]^\ell$ ($\delta > 0$) such that for all $\vec{p} \in [0, \delta]^\ell$, the limiting vector satisfies $\vec{0} < \vec{z}_\infty < \vec{1}$. In order to do so, define $d = \max_{i=0}^{\ell-1} d_i$ and $\theta = \min_{i=0}^{\ell-1} \theta_i \geqslant 2$ and introduce the function $\phi(x;p) : [0,1] \to [0,1]$ given by

$$\phi(x;p) := p + (1-p) \sum_{k=\theta}^{d} \binom{d}{k} x^k (1-x)^{d-k}. \tag{14}$$

In view of Lemma 1, there exists $p_c \in (0,1)$ such that 1 is the only solution in $[0,1]$ of $\phi(x;p) = x$ for all $p > p_c$. On the other hand there exist two solutions in $(0,1)$ for $p < p_c$, and one if $p = p_c$. From stochastic dominance it follows that

$$\mathbb{P}\left(\mathrm{Bin}\,(d_i, p_i) \geqslant \theta_i\right) \leqslant \mathbb{P}\left(\mathrm{Bin}\,(d, p_i) \geqslant \theta\right), \tag{15}$$

so

$$\phi_i(x; p_i) \leqslant \phi(x; p_i) \tag{16}$$

for all x. Choose $0 < \delta < p_c$ and consider the following mapping with $\vec{Z}_0 = \delta$:

$$\vec{Z}_t = \phi(\vec{Z}_{t-1}; \delta). \tag{17}$$

The limit $\vec{Z}_\infty := \lim_{t \to \infty} \vec{Z}_t$ exists and $\vec{Z}_\infty < 1$ by the choice of δ, cf. [14]. Now choosing all $p_i \leqslant \delta$, it inductively follows from (16) and (17) that

$$\vec{z}_{i,t} = \phi_i(\vec{z}_{i+1,t-1}; p_i) \leqslant \phi(\vec{z}_{i+1,t-1}; \delta) \leqslant \phi(\vec{Z}_{t-1}; \delta) = \vec{Z}_t. \tag{18}$$

Hence $\vec{z}_{i,t} \leqslant \vec{Z}_t$ for every i and t, and $\vec{z}_{i,\infty} \leqslant \vec{Z}_\infty < 1$ for all i. This concludes the first part of the proof and shows that $[0, \delta]^\ell \subset W_0$.

By definition, for every $\vec{p} \notin W_0$ it follows that $\vec{z}_{i,\infty} = 1$ for some i, hence $\vec{z}_\infty = \vec{1}$, i.e. $\mathbb{T}$ a.a.s. fully percolates for all $\vec{p} \in \overline{W}_0$. This proves statement (ii).

Finally, we want to show that W_0 is contained in some hypercube of volume σ^ℓ, where $\sigma > 0$. In order to do so, for every $\vec{p}_{-i} \in [0,1]^{\ell-1}$, define the critical value $h_c(\vec{p}_{-i})$ as the infimum of the probability p_i necessary such that $\mathbb{T}$ fully percolates a.a.s.:

$$h_c(\vec{p}_{-i}) = \inf \left\{ s : \mathbb{T} \text{ a.a.s. fully percolates for probabilities } \vec{p} = (\vec{p}_{-i}|s) \right\}. \tag{19}$$

Taking $p_i = 1$ will yield $\vec{z}_\infty = \vec{1}$, so the critical value $h_c(\vec{p}_{-i})$ is well defined. Next we want to show that h_c is not trivially identical to 1 on the entire domain $[0,1]^{\ell-1}$.

Lemma 2. *There exists a constant* $\sigma \in (0,1)$ *such that for every vector of initial probabilities* $\vec{p} \in [0,1]^\ell$ *and every coordinate* $i \in \{0, 1, \ldots, \ell-1\}$, *the threshold function satisfies* $h_c(\vec{p}_{-i}) \leqslant \sigma$.

Proof. We have $\varphi_{d,p,\theta}(x) \geqslant \varphi_{d,p,d-1}(x) \geqslant \varphi_{d,0,d-1}(x)$ for every $x \in [0,1]$, so for all $i \in \{0, 1, \ldots, \ell-1\}$,

$$\varphi_{d_i,p_i,\theta_i}(x) \geqslant \varphi_{d_i,p_i,d_i-1}(x) \geqslant \varphi_{d_i,0,d_i-1}(x) = d_i x^{d_i-1} - (d_i - 1)x^{d_i}. \tag{20}$$

It is easy to show that $d_i x^{d_i-1} - (d_i - 1)x^{d_i} = x$ has always one real solution in $(0,1)$, call it s_i, and that $d_i x^{d_i-1} - (d_i - 1)x^{d_i} > x$ for $s_i < x < 1$. Without loss of generality, let s_0 be the maximum among all s_i. Choose any σ such that $s_0 < \sigma < 1$, and consider the recurrence system given by:

$$\vec{u}_{0,t} = \varphi_{d_0,\sigma,d_0-1}(\vec{u}_{1,t-1}),$$
$$\vec{u}_{i,t} = \varphi_{d_i,0,d_i-1}(\vec{u}_{i+1,t-1}),$$

$i = 1, 2, \ldots, \ell - 1$, with initial value $\vec{u}_0 = (\vec{u}_{0,0}, \vec{u}_{1,0}, \ldots, \vec{u}_{\ell-1,0}) = (\sigma, 0, \ldots, 0)$. The limit $\vec{u}_\infty := \lim_{t\to\infty} (\vec{u}_{0,t}, \vec{u}_{1,t}, \ldots, \vec{u}_{\ell-1,t})$ exists by the monotone convergence theorem. Moreover, $\vec{u}_{i,\infty} \geqslant \sigma$ for all i by the choice of σ, which in turn implies $\vec{u}_{i,\infty} = 1$ for all i (using the aforementioned fact that $\varphi_{d_i,p,d_i-1}(x) \geqslant d_i x^{d_i-1} - (d_i - 1)x^{d_i} > x$ for $s_i < x < 1$). For every initial vector $\vec{p}$ for which $p_0 \geqslant \sigma$, it follows from (20) that $\vec{z}_{i,t} \geqslant \vec{u}_{i,t}$ for all i, t, hence $\vec{z}_\infty = \vec{1}$. Thus, for every $\vec{p}$, we have $h_c(\vec{p}_0) \leqslant \sigma < 1$. In the same way, it follows that $h_c(\vec{p}_{-i}) \leqslant \sigma < 1$ for all i.

If $\vec{p} \notin [0,\sigma]^\ell$, then $p_i > \sigma \geqslant h_b(\vec{p}_{-i})$ for at least one i by Lemma 2. Hence $\vec{\mathbb{T}}$ a.a.s. fully percolates by definition of h_c. This means that $W_0 \subset [0,\sigma]^\ell$, which concludes the proof of Theorem 1.

3.3   Region of Full Percolation

In the following lemma we provide better bounds on p_i for full percolation.

Lemma 3. $\vec{\mathbb{T}}$ *a.a.s. fully percolates for any initial vector of probabilities* $\vec{p}$ *such that for all* $i \in \{0, 1, \ldots, \ell - 1\}$, $p_i \in (1 - 1/\beta_i, 1]$, *where*

$$\beta_i := d_i \left(\frac{d_i - 1}{\theta_i - 1}\right) \left(\frac{\theta_i - 1}{d_i - 1}\right)^{\theta_i - 1} \left(\frac{d_i - \theta_i}{d_i - 1}\right)^{d_i - \theta_i}. \tag{21}$$

We remark that $\beta_i \geqslant 1$, as will be shown below.

Proof. Consider again the function $F_i(z)$, given by (13). The first derivative of $F_i(z)$ is

$$F_i'(z) = \prod_{j=i}^{i-1 \bmod \ell} \phi_j'\left(\phi_{j+1}(\ldots \phi_{i-1}(z; p_{i-1}) \ldots p_{j+1}); p_j\right). \tag{22}$$

For every $\phi_i(x; p_i)$, the first derivative is given by:

$$\phi_i'(x; p_i) = (1 - p_i)d_i \left(\frac{d_i - 1}{\theta_i - 1}\right) x^{\theta_i - 1}(1 - x)^{d_i - \theta_i}, \tag{23}$$

and by differentiating again one finds that the maximum of $\phi_i'(x; p_i)$ is attained at $(\theta_i - 1)/(d_i - 1)$:

$$\max_{x \in [0,1]} \phi_i'(x; p_i) = \phi_i'\left(\frac{\theta_i - 1}{d_i - 1}; p_i\right) = (1 - p_i)\beta_i \tag{24}$$

by the definition of β_i. Note that β_i is the maximum of $B_i'(x)$, where B_i is given by (7). Since $B_i(0) = 0$ and $B_i(1) = 1$, it follows from the mean value theorem that $\beta_i \geqslant 1$.

For $p_i \in (0, 1)$, the first derivative given in (23) is strictly greater than 0, i.e. $\phi_i'(x; p_i) > 0$. For $p_i > 1 - 1/\beta_i$, the maximum of the first derivative is strictly less than 1, i.e. $\max_{0 \leqslant x \leqslant 1} \phi_i'(x; p_i) < 1$, see (24). Hence, for any vector of probabilities that satisfies $p_i > 1 - 1/\beta_i$ for all i, the convolution given in (22) yields

$$F_i'(z) - 1 < 0, \tag{25}$$

for all $i \in \{0, 1, \ldots, \ell - 1\}$. Thus the first derivative of the equation $F_i(z) - z$ is strictly negative on $[0, 1]$. Moreover $F(0) > 0$ and $F(1) - 1 = 0$, hence $z = 1$ is the only solution of $F_i(z) = z$ on $[0, 1]$. This implies that $\vec{z}_\infty = \vec{1}$ for any $\vec{p}$ that satisfies the condition of the lemma. Also note that if at least one p_i is 1, i.e. $\vec{z}_{i,0} = 1$, then $\vec{z}_{i,\infty} = 1$ for all i in view of (5), which concludes the proof.

3.4 Trajectory of $\vec{z}_t$

In this section we analyze the trajectory of $\vec{z}_t$ over time $t = 0, 1, \ldots$ More precisely, we show necessary and sufficient conditions on $\vec{z}_t$ such that the initial vector $\vec{p}$ lies in W_0.

To start, consider again a function of the form $\phi(x; s)$ defined by (cf. (3))

$$\phi(x; s) := s + (1 - s) \sum_{k=\theta}^{d} \binom{d}{k} x^k (1 - x)^{d-k} \tag{26}$$

for certain parameters d and θ. Let $L(s) \leqslant R(s)$ be the real solutions of $\phi(x; s) = x$ in $(0, 1)$ if such solutions exist. We know that there exists some critical $s_c \in (0, 1)$, such that: (i) if $s < s_c$, there are two real solutions $L(s) < R(s)$ in $(0, 1)$; (ii) if $s = s_c$, there is one solution $L(s) = R(s)$ in $(0, 1)$; (iii) if $s > s_c$ there are no real solutions in $(0, 1)$, see Lemma 1. It is easy to show the following.

Lemma 4. *The limit of the sequence defined by the iteration $x_{t+1} := \phi(x_t; p)$ for $t = 0, 1, \ldots$ satisfies*

$$\lim_{t \to \infty} x_t = \begin{cases} L(p), & x_0 \in [0, R(p)) \text{ and } p \leqslant p_c, \\ R(p), & x_0 = R(p) \text{ and } p \leqslant p_c, \\ 1, & \text{otherwise.} \end{cases}$$

We will write L_i and R_i for the functions of Lemma 4 associated with ϕ_i.

Lemma 5. *We have $\vec{z}_\infty < \vec{1}$ if and only if $\vec{z}_{i,t} \leqslant R_{i-1}(p_{i-1})$ for every i and every t.*

Proof. First, $\vec{z}_{i,t} = \phi_i(\vec{z}_{i+1,t-1}; p_i)$. Iterating this equation ℓ times it follows that

$$\vec{z}_{i,t+\ell} = \phi_i\left(\phi_{i+1}\left(\cdots\left(\phi_{i-1}\left(\vec{z}_{i,t}; p_{i-1}\right)\cdots\right); p_{i+1}\right); p_i\right). \tag{27}$$

Assume that there exists some i such that $\vec{z}_{i,t} > R_{i-1}(p_{i-1})$. Let $b_0 := \vec{z}_{i,t}$ and define $b_k := \phi_{i-1}(b_{k-1}; p_{i-1})$ for $k \geqslant 1$. The composition

$$\phi_i \circ \phi_{i+1} \circ \cdots \phi_{\ell-1} \circ \phi_0 \circ \cdots \circ \phi_{i-2}$$

is increasing, as the convolution of increasing functions, hence from (27) we obtain $\vec{z}_{i,t+\ell \cdot k} \geqslant b_k$. From Lemma 4 it follows that $\lim_{k \to \infty} b_k = 1$, so $\vec{z}_{i,\infty} = 1$ and consequently $\vec{z}_{j,\infty} = 1$ for all j. Conversely, if $\vec{z}_{i,t} \leqslant R_{i-1}(p_{i-1})$ for every i and every t, then $\vec{z}_{i,\infty} \leqslant R_{i-1}(p_{i-1}) < 1$ for all i and thus $\vec{z}_\infty < \vec{1}$.

Lemma 6. *Let a be the index for which $R_j(p_j)$ is maximal, and set $R_{\max} := R_a(p_a)$. We have $\vec{z}_\infty < \vec{1}$ if and only if $\vec{z}_{i,t} \leqslant R_{\max}$ for every i and t.*

Proof. First, let us recall the following two facts: (1) $\vec{z}_{a,\infty} < 1$ if and only if $\vec{z}_\infty < \vec{1}$; (2) $\vec{z}_{a,\infty} = 1$ if and only if $\vec{z}_\infty = \vec{1}$.

To prove sufficiency, assume that there exist i and t such that $\vec{z}_{i,t} > R_{\max}$. Then by Lemma 4 it follows that $\vec{z}_{i,\infty} = 1$ for every i, thus $\vec{z}_\infty = \vec{1}$.

To prove necessity, let $\vec{z}_{i,t} \leqslant R_{\max}$ for all i and t. It follows immediately that $\vec{z}_{i,\infty} \leqslant R_{\max} < 1$ for all i, completing the proof.

3.5   Bootstrap Percolation on an Unoriented Tree $\mathbb{T}$

To determine the critical region for bootstrap percolation on $\mathbb{T}$, we use the result of Sect. 3.1 on oriented trees, as in [14]. Let z_t be the probability that the root is active at time t, and define the limiting probability $z_\infty := \lim_{t \to \infty} z_t$.

Theorem 2. *The probability z_∞ is given by*

$$z_\infty = p_0 + (1 - p_0) \sum_{k=\theta_0}^{d_0+1} \binom{d_0 + 1}{k} \vec{z}_{1,\infty}^{*k} (1 - \vec{z}_{1,\infty})^{d_0+1-k}. \tag{28}$$

Proof. As before, p_0 simply stands for the probability that the root is initially active, so we focus on the case that it is initially inactive, which happens with probability $1 - p_0$. In this case, it can become active if at least θ_0 of its $d_0 + 1$ neighbors become active in the process. For the root activity, it is immaterial whether or not a node can contribute to activating neighboring nodes that are further away from the root, so we can consider the d_0+1 root branches as oriented trees (oriented towards the root) on which oriented bootstrap percolation is performed. Thus we know that $\vec{z}_{1,\infty}$ is the limiting probability for a root neighbor to become active (if the root is not initially), which proves the desired formula.

Theorem 3. *The percolation regions on oriented tree $\vec{\mathbb{T}}$ and unoriented tree $\mathbb{T}$ are the same and equal to $\overline{W}_0$.*

Proof. If the unoriented tree fully percolates a.a.s., then in particular $z_\infty = 1$. Note that $z_\infty = 1$ by Theorem 2 if and only if $\vec{z}_{1,\infty} = 1$ (the case $p_0 = 1$ is trivial). However, if $\vec{z}_{1,\infty} = 1$, then also $\vec{z}_{i,\infty} = 1$ for all i, which means that even the oriented tree percolates a.a.s. The converse is clear as well.

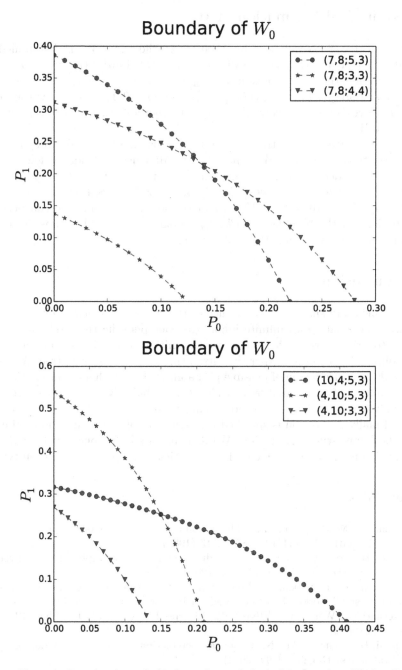

Fig. 1. Numerical evaluation of the boundary of the critical region W_0 in dimension two for different values of degrees and activation thresholds. Concretely, $(d_0, d_1; \theta_0, \theta_1)$ takes the values $(7, 8; 5, 3), (7, 8; 3, 3), (7, 8; 4, 4)$ in the upper and $(10, 4; 5, 3), (4, 10; 5, 3), (4, 10; 3, 3)$ in the lower diagram.

4 Numerical Estimation of W_0

For a given tree $\vec{\mathbb{T}}$ (as well as $\mathbb{T}$ by Theorem 2), the region W_0 is determined in Theorem 1. That is, W_0 is the set of the initial probabilities $(p_0, p_1, \ldots, p_{\ell-1}) \in [0, 1]^\ell$ such that there exists a solution $(x_0, x_1, \ldots, x_{\ell-1}) \in [0, 1)^\ell$ (note: all $x_i < 1$) of the system given by $x_i = \phi_i(x_{i+1}; p_i)$, where $i = 0, 1, \ldots, \ell - 1$. At the same time, this gives a criterion how one can decide whether $(p_0, p_1, \ldots, p_{\ell-1}) \in [0, 1]^\ell$ belongs to W_0.

In Fig. 1, we demonstrate results of this method. We present the boundary that separates W_0 and $\overline{W}_0$ in dimension 2 for a few different values of degrees d_0, d_1 and activation thresholds θ_0, θ_1. Specifically, $(d_0, d_1; \theta_0, \theta_1)$ takes the values $(7, 8; 5, 3), (7, 8; 3, 3), (7, 8; 4, 4)$ in the upper and $(10, 4; 5, 3), (4, 10; 5, 3), (4, 10; 3, 3)$ in the lower diagram. One can observe monotonicity of the boundary in (θ_0, θ_1). Looking at these diagrams, it is also tempting to conjecture that W_0 is always convex.

5 Conclusion

We examined and showed the existence of the region of critical probabilities in bootstrap percolation on infinite inhomogeneous periodic trees. The main difference to prior work is that we allow the initial probabilities and activation thresholds to be periodic, rather than fixed throughout the entire tree. We characterized the entire region of possible probabilities for which all nodes in the tree become a.a.s. active, as well as provided the probability that a node becomes eventually active, for any initial probability vector that does not belong to this region. Finally, the region is specified through a set of equations whose solution gives the boundary of the region. We demonstrated how one can numerically find this boundary and provided a few numerical examples in dimension two.

References

1. Aizenman, M., Lebowitz, J.L.: Metastability effects in bootstrap percolation. J. Phys. A: Math. Gen. **21**(19), 3801–3813 (1988)
2. Amini, H.: Bootstrap percolation and diffusion in random graphs with given vertex degrees. Electron. J. Comb. **17**, 1–20 (2010). #R25
3. Amini, H., Fountoulakis, N.: What I tell you three times is true: bootstrap percolation in small worlds. In: Proceedings of Internet and Network Economics - 8th International Workshop, WINE 2012, Liverpool, UK, 10–12 December 2012, pp. 462–474 (2012)
4. Amini, H., Fountoulakis, N.: Bootstrap percolation in power-law random graphs. J. Stat. Phys. **155**(1), 72–92 (2014)
5. Amini, H., Fountoulakis, N., Panagiotou, K.: Bootstrap percolation in inhomogeneous random graphs. arXiv:1402.2815
6. Balogh, J., Bollobás, B., Duminil-Copin, H., Morris, R.: The sharp threshold for bootstrap percolation in all dimensions. Trans. Am. Math. Soc. **364**(5), 2667–2701 (2012)

7. Balogh, J., Bollobás, B., Morris, R.: Majority bootstrap percolation on the hypercube. Comb. Probab. Comput. **18**(1–2), 17–51 (2009)
8. Balogh, J., Bollobás, B., Morris, R.: Bootstrap percolation in high dimensions. Comb. Probab. Comput. **19**(5–6), 643–692 (2010)
9. Balogh, J., Peres, Y., Pete, G.: Bootstrap percolation on infinite trees and nonamenable groups. Comb. Probab. Comput. **15**(5), 715–730 (2006)
10. Balogh, J., Pittel, B.: Bootstrap percolation on the random regular graph. Random Struct. Algorithms **30**(1–2), 257–286 (2007)
11. Biskup, M., Schonmann, R.H.: Metastable behavior for bootstrap percolation on regular trees. J. Stat. Phys. **136**(4), 667–676 (2009)
12. Bollobás, B., Gunderson, K., Holmgren, C., Janson, S., Przykucki, M.: Bootstrap percolation on Galton-Watson trees. Electron. J. Probab. **19**(13), 1–27 (2014)
13. Bradonjić, M., Saniee, I.: Bootstrap percolation on random geometric graphs. Probab. Eng. Inf. Sci. **28**(2), 169–181 (2014)
14. Bradonjić, M., Saniee, I.: Bootstrap percolation on periodic trees. In: Proceedings of 12th Workshop on Analytic Algorithmics and Combinatorics, ANALCO 2015, San Diego, CA, USA, 4 January 2015, pp. 89–96 (2015)
15. Chalupa, J., Leath, P.L., Reich, G.R.: Bootstrap percolation on a Bethe lattice. J. Phys. C **12**, L31 (1979)
16. Duminil-Copin, H., Van Enter, A.C.D.: Sharp metastability threshold for an anisotropic bootstrap percolation model. Ann. Probab. **41**(3A), 1218–1242 (2013)
17. Fontes, L., Schonmann, R.: Bootstrap percolation on homogeneous trees has 2 phase transitions. J. Stat. Phys. **132**(5), 839–861 (2008)
18. Gravner, J., Holroyd, A.E., Morris, R.: A sharper threshold for bootstrap percolation in two dimensions. Probab. Theor. Relat. Fields **153**(1–2), 1–23 (2012)
19. Holroyd, A.E.: Sharp metastability threshold for two-dimensional bootstrap percolation. Probab. Theor. Relat. Fields **125**, 195–224 (2003)
20. Janson, S., Luczak, T., Turova, T., Vallier, T.: Bootstrap percolation on the random graph $G_{n,p}$. Ann. Appl. Probab **22**(5), 1989–2047 (2012)
21. Schonmann, R.: Critical points of two-dimensional bootstrap percolation-like cellular automata. J. Stat. Phys. **58**(5–6), 1239–1244 (1990)
22. Schonmann, R.H.: On the behavior of some cellular automata related to bootstrap percolation. Ann. Probab. **20**(1), 174–193 (1992)
23. van Enter, A., Adler, J., Duarte, J.: Finite-size effects for some bootstrap percolation models. J. Stat. Phys. **60**(3–4), 323–332 (1990)
24. van Enter, A., Fey, A.: Metastability thresholds for anisotropic bootstrap percolation in three dimensions. J. Stat. Phys. **147**(1), 97–112 (2012)
25. van Enter, A., Hulshof, T.: Finite-size effects for anisotropic bootstrap percolation: logarithmic corrections. J. Stat. Phys. **128**(6), 1383–1389 (2007)

Fast Approximation Algorithms for p-centers in Large δ-hyperbolic Graphs

Katherine Edwards[1]([✉]), Sean Kennedy[2], and Iraj Saniee[2]

[1] Department of Computer Science, Princeton University, Princeton, NJ 08540, USA
katherine.edwards2@gmail.com
[2] Mathematics of Networks and Communications Department,
Bell Labs, Nokia, Holmdel, NJ 07974, USA
{kennedy,iis}@research.bell-labs.com

Abstract. We provide a quasilinear time algorithm for the p-center problem with an additive error less than or equal to 3 times the input graph's hyperbolic constant. Specifically, for the graph $G = (V, E)$ with n vertices, m edges and hyperbolic constant δ, we construct an algorithm for p-centers in time $O(p(\delta+1)(n+m)\log_2(n))$ with radius not exceeding $r_p + \delta$ when $p \leq 2$ and $r_p + 3\delta$ when $p \geq 3$, where r_p are the optimal radii. Prior work identified p-centers with accuracy $r_p + \delta$ but with time complexity $O((n^3 \log_2 n + n^2 m) \log_2(\mathrm{diam}(G)))$ which is impractical for large graphs.

1 Introduction

The p-center algorithm is a discrete variant of arguably one of the most frequently used clustering algorithms, the k-means clustering. The goal of the p-center algorithm is to identify on a given graph a pre-specified number p of vertices or centers, such that the maximum distance of any graph vertex to its nearest p-center is minimized. For any given p, the algorithm naturally partitions a graph into p clusters induced by the position of its p-centers. Clusters induced by the p-centers are not necessarily balanced as these are determined strictly by the metric properties of the graph. Thus p-center clustering is more appropriate for distance-based partitioning or classification than other frameworks, such as community detection. Unfortunately, as a clustering algorithm the complexity of the p-center algorithm is generally prohibitive, $O(n^p)$ for an n-node graph, making it inapplicable to even moderate size graphs.

Proved nearly four decades ago, Shier's minimax result for trees and metric trees [15] leads to an exact algorithm with quasilinear time complexity (in the number of vertices and edges of the graph) for determination of an optimal set of p-centers by repeatedly finding diagonal pairs on the graph and carving out a ball containing one end of the current diagonal pair. Hochbaum and Shmoys [10] give a (multiplicative) 2-approximation algorithm for determining p-centers in graphs satisfying the triangle inequality with running time $O(m \log_2 m)$. Subsequently, Dyer and Frieze [4] improve this to a 2-approximation algorithm with running time $O(np)$. These algorithms are, in a sense, best possible as Hsu and

© Springer International Publishing AG 2016
A. Bonato et al. (Eds.): WAW 2016, LNCS 10088, pp. 60–73, 2016.
DOI: 10.1007/978-3-319-49787-7_6

Nemhauser [11] show that determining an α-approximate solution to p-centers is NP-hard whenever $\alpha < 2$.

In an insightful paper [3], Chepoi and Estellon essentially apply the technique of Shier [15] to graphs with small hyperbolic constant, δ. These are graphs whose metric structure differs from the metric structure of a tree by a fixed constant (as explained in Sect. 2 and, in particular, Sect. 2.2 and Fig. 1. For more details see [1,3,9]). The algorithmic version of this scheme [2] gives rise to what is essentially an $O(n^3)$ approximation for p-center on an n-vertex graph with hyperbolic constant δ appearing both as a prefactor in the complexity expression and also in the degree of approximation in terms of an additive constant to the radius of the optimal p-center partition. Of course the polynomial time complexity $O(n^3)$ is still impractical for graphs of hundreds of thousands to millions of nodes as would be even a quadratic complexity approximation.

Since there is evidence that real-life networks extracted from social media, co-authorship and collaboration, friendship and many other settings, have small hyperbolic constants [13], it would be desirable to know if the cubic complexity is tight or can be further reduced, at least by negotiating on the degree of the approximation. In this paper we show that by giving up to 3δ in the (additive) approximation, one can achieve a quasilinear time p-center approximation. As such, this scheme is the first p-center approximation applicable to large graphs, particularly when p is relatively small, for example in the range 10–10^4 and n is large, for example, 10^5–10^9 vertices.

In the following sections we describe how the cubic complexity of [3] to quasilinear reduction is achieved without adding more than 3δ to the radius of the optimal p-center clusters. In Sect. 2 we outline necessary definitions, in particular, for geodesic metric spaces (Sect. 2.1) and hyperbolicity (Sect. 2.2). We then turn to a more formal discussion of p-centers, p-packings, and the dual problems which take center stage in our discussion (Sect. 3). In Sect. 3.1 we focus on algorithms for solving and approximating these problems on δ-hyperbolic graphs. The formal statements of our main results are also found in Sect. 3.1. Section 4 contains the ingredients for the proofs of the main results. Due to space constraints, the details of these proofs are omitted and can be found in the full version of this paper [5]. We finish in Sect. 5 with experimental validation of our algorithms. The reader is invited to consult the full version of this paper which contains all the details missing from this abstract [5].

2 Definitions and Notation

Let $G = (V, E)$ be an undirected graph, with V the set of vertices and E the set of edges. To each edge uv, we associate a line segment of length 1, so that we may refer to any point on uv at distance t from u and $1 - t$ from v ($0 \le t \le 1$). This (uncountably infinite) set of points of G is denoted $A(G)$. We will use the notation $n = |V(G)|$ and $m = |E(G)|$. In this paper, the distance $d(u, v)$ between any two points u and v in $A(G)$ is the length of a shortest path between them in G. When u and v are vertices, we write $[u, v]$ to refer to a shortest (also called *geodesic*) path.

Note that shortest paths need not be unique. For a geodesic path $P = [u, v]$ and $i \in [0, d(u, v)]$, the point $P[i]$ is the one at distance i from u on P.

2.1   Geodesic Metric Spaces and Graphs

Let (X, d) be a metric space. If x, y are points in X, a *geodesic segment* $[x, y]$, when it exists, is a continuous curve parametrized by a line segment of length $d = d(x, y)$. That is, a map $\rho : [0, d] \to X$ with $\rho(0) = x$, $\rho(d) = y$ and $d(\rho(s), \rho(t)) = |s - t|$ for each $s, t \in [0, d]$. A metric space is *geodesic* if there exists a geodesic segment joining every pair of points. Note that geodesic segments need not be unique, e.g. a diagonal pair of points on a cycle.

Any graph as we have defined above can be viewed as a geodesic metric space $(A(G), d)$. Such a metric space is called *graphic* and it will be convenient in what follows to think of graphs in this way. In a graphic metric space, a geodesic $[x, y]$ is simply a shortest path from x to y regardless of x and y being in $V(G)$ or in $A(G)$.

Let $S \subseteq X$ be compact. The *diameter* diam(S) of is the maximum length of a geodesic between two vertices in S. For $u \in S$, $F_S(u)$ is the set of points in S whose distance from u is maximum. Two points $u, v \in S$ are *diametrical* if $d(u, v) = $ diam(S). They are *locally diametrical* if $u \in F_S(v)$ and $v \in F_S(u)$. It follows that $d(u, v \in F_S(u)) \leq$ diam(S) and $d(v, u \in F_S(v)) \leq$ diam(S).

If v is a point of $A(G)$ and $r \in \mathbb{R}$, we write $B_r(v)$ for the closed ball of radius r about v, i.e. all points at distance at most r from v. For a geodesic path $P = [u, v]$ and for the length $0 \leq \theta < d(u, v)$, the point $i = [u, v][\theta] \in A(G)$ is at distance θ from u on P. When there is no ambiguity, we identify the point $i = P[\theta]$ with the length θ. Clearly the two points $[u, v][i]$ and $[v, u][i]$ do not generally coincide.

2.2   Hyperbolicity

The concept of hyperbolicity of a metric space was introduced by Rips and Gromov in [9]. There are several essentially equivalent definitions but in this paper we will mainly use the δ-*thin-triangle* characterization.[1] For points x, y, z in X, we write $\Delta(x, y, z)$ to denote a *geodesic triangle* formed by x, y, z; that is the union of three geodesics $[x, y], [y, z], [x, z]$ (usually the choice of geodesics won't matter).

Given a geodesic triangle $\Delta \equiv \Delta(x, y, z)$, let π be half the perimeter, $\pi = \frac{1}{2}(d(x, y) + d(y, z) + d(x, z))$ and define $\alpha_x = \pi - d(y, z)$ and similarly $\alpha_y = \pi - d(x, z)$ and $\alpha_z = \pi - d(x, y)$. Thus $\alpha_x + \alpha_y = d(x, y)$ and so on. One can imagine a triangle drawn in the Euclidean plane with side lengths $d(x, y), d(x, z)$ and $d(y, z)$. Its inscribed circle would touch the triangle sides $[x, y], [y, z]$ and $[z, x]$ at points m_z, m_x and m_y respectively. From elementary geometry, $[x, y][\alpha_x] = [y, x][\alpha_y] = m_z$ and $[y, z][\alpha_y] = [z, y][\alpha_z] = m_x$ and $[z, x][\alpha_z] = [x, z][\alpha_x] = m_y$, as illustrated in Fig. 1.

---

[1] For a comprehensive treatment of δ-hyperbolicity see [1].

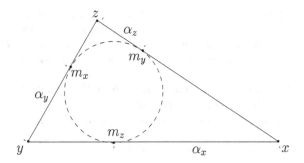

Fig. 1. A geodesic triangle $\Delta(x, y, z)$ with internal points m_x, m_y and m_z and internal distances α_x, α_y and α_z labelled.

The points m_x, m_y, m_z are called the *internal points* and $\alpha_x, \alpha_y, \alpha_z$ the *internal distances* corresponding to x, y, z respectively in Δ. The *insize* of the triangle Δ is the maximum of $\max_{\theta \in [0,\alpha_x]} d([x, y][\theta], [x, z][\theta])$, $\max_{\theta \in [0,\alpha_y]} d([y, x][\theta], [y, z][\theta])$, and $\max_{\theta \in [0,\alpha_z]} d([z, x][\theta], [z, y][\theta])$.

Definition 1. Let (X, d) be a geodesic metric space, and $\delta \geq 0$. X is δ-hyperbolic (equivalently, the hyperbolicity of X is δ) if the insize of every geodesic triangle is at most δ. Let δ be minimum such that the insize of every geodesic triangle is at most δ. We say that X is δ-hyperbolic (equivalently, the hyperbolicity of X is δ).

If G is a graph whose associated graphic metric space is δ-hyperbolic then we say G is δ-hyperbolic. The reader may verify that every tree is 0-hyperbolic. Hyperbolicity is sometimes defined in terms of a four-point condition.

Lemma 1 (4-point condition, see Proposition 1.22 in [1]). *Let (X, d) be a δ-hyperbolic metric space. There is a constant $\delta_{4-point} \leq \delta$ such that for any 4 points $x, y, z, w \in X$, their ordered set of sums of opposite sides, wlog $d(x, y) + d(w, z) \geq d(x, z) + d(y, w) \geq d(x, w) + d(y, z)$, satisfy $d(x, y) + d(w, z) - d(x, z) - d(y, w) \leq 2\delta_{4-point}$.*

The fact that in a δ-hyperbolic metric space $\delta_{4-point}$ is always less than or equal to δ follows directly from the proof of Proposition 1.22 on page 411.

3 p-centers and p-packings

Let (X, d) be a geodesic metric space and S be a compact subset of X. Throughout this paper we rely on two intimately related notions, p-centers and p-packings.

Definition 2 (p-centers). A set $C \subset X$ r-*dominates* S if for every point $s \in S$ there exists a point $c \in C$ with $d(s, c) \leq r$. The p-*radius* of S, denoted by $r_p(S)$, is the minimum r such that there exists a set of at most p points $C_p(S)$ that r-dominates S. The points in $C_p(S)$ are called p-*centers* of S.

Definition 3 (p-packings). A set $D \subseteq S$ is an r-*dispersion* in S if each pair of points $s, s' \in D$, $s \neq s'$, $d(s, s') \geq r$. The p-*diameter* of S, denoted by $d_p(S)$, is the maximum r such that there exists a set of at least p points $D_p(S)$ that is an r-dispersion in S. The points in $D_p(S)$ are called a p-*packing*.

Consider a set of p points C which r-dominate S. By definition, for any choice of $p + 1$ points D, each $d \in D$ is within r of some $c \in C$, and by the pigeonhole principle, at least two, say a_1 and a_2, are within r of the same $c \in C$. Hence,

$$d(a_1, a_2) \leq d(a_1, c) + d(a_2, c) \leq 2r.$$

So, $\min_{i \neq j} d(a_i, a_j) \leq 2r$. Since this holds for all choices of C and D, we have the following observation which first appeared in [15].

Observation 4. $r_p(S) \geq \frac{1}{2} d_{p+1}(S)$.

It turns out that these two invariants are equal whenever S has a tree-metric. Indeed, Shier showed the following.

Theorem 5 (Shier [15]). *Let T be a tree. Then* $r_p(T) = \frac{1}{2} d_{p+1}(T)$.

As discussed in Sect. 2.2, δ-hyperbolic spaces are treelike, by which we mean that they possess a metric structure that differs from a tree metric by δ. Therefore, it is logical to attempt to extend Shier's result on p-center covering and packing to such structures. Chepoi and Estellon [3] do exactly this by giving an elegant extension of Shier's theorem to δ-hyperbolic spaces.

Theorem 6 (Chepoi and Estellon [3]). *Let X be a δ-hyperbolic metric space and S a finite subset of X. Then*

$$r_p(S) \leq \frac{1}{2} d_{p+1}(S) + \delta$$

This relationship between $r_p(S)$ and $d_{p+1}(S)$ is a key element in algorithms for approximating p-centers and p-packing.

3.1 Algorithms for p-centers and p-packings

The p-packing problem, sometimes referred to as the p-dispersion problem, has received some attention in the literature. For example it is known to be NP-hard [6]. Highly relevant to our work is the heuristic that iteratively adds each of the p points by maximizing the points' distance from previously chosen points (see for example [7,14]). This heuristic is shown to be a 2-approximation algorithm by Ravi et al. [14]. For more information, we refer the interested reader to [8] that has an empirical comparison of ten p-dispersion heuristics.

To our knowledge, the previous best algorithm in terms of an additive error not exceeding δ for the p-radius follows from the Chepoi-Estellon bound (Theorem 6). Indeed, the proof in [3] leads to a polynomial algorithm to solve

p-centers in graphs with an additive error of δ on the p-radius.[2] Specifically, in time $O((n^3 \log_2 n + n^2 m) \log_2(\mathrm{diam}(G)))$ the authors in [3] determine a set U of p points such that U $(r_p + \delta)$-dominates $V(G)$. Their algorithm involves finding diametrical pairs of vertices in subsets of $V(G)$ $O(n \log_2(\mathrm{diam}(G)))$ times. Johnson's algorithm [12] finds the diameter in time $O(n^2 \log_2 n + nm)$; hence the running time in Chepoi-Estellon [3] follows.

As pointed out in the introduction, in this work we leverage the fact that instead of finding diametrical pairs, one can just use **locally** diametrical pairs (introduced in Sect. 2.1) with significant reduction in computational time with only a small penalty in the p-radius. Our main result is the following.

Theorem 7. *Let G be a δ-hyperbolic graph, $p \geq 3$ an integer and $r_p(G)$ the optimal radius of the p-center for $V(G)$. There exists an algorithm to find a set of p points that $(r_p + 3\delta)$-dominates $V(G)$. Further, the algorithm runs in time $O(n \log_2 n + (m+n)((2p+1)(\lceil 4 + 3\delta + 2\delta \log_2 n \rceil) + (p+1))) = O(p(\delta+1)(m + n) \log_2 n)$.*

Though the Chepoi-Estellon algorithm [3] achieves a better approximation (an additive factor of δ instead of our 3δ), its running time is $O((n^3 \log_2 n + n^2 m) \log_2(\mathrm{diam}(G)))$. We first show below how to improve their running time by a factor of n (Lemma 3), but this approach still remains infeasible for large graphs. When $p \in \{1, 2\}$ we can achieve the same Chepoi-Estellon p-radius bound but in quasilinear time.

Theorem 8. *Let (X, d) be a δ-hyperbolic metric space, S a finite subset of X and $p \in \{1, 2\}$. There exists an algorithm to determine a set of p points that $(r_p + \delta)$-dominate S. Further, the algorithm runs in time $O((2\delta + 1)t_X)$, where t_X is the time required to find the set of points at maximum distance from a given point in X. In particular in a δ-hyperbolic graph the running time is $O((2\delta + 1)(m + n))$.*

For $p = 1$, the previous best algorithm we know of is due to Chepoi et al. [2]: the approximation error is $\leq 5\delta$, and the computation requires just two breadth-first searches. In contrast, we require $2\delta + 1$ breadth-first searches to achieve the smaller additive factor of δ.

The remainder of this section is organized as follows. We start by showing how to improve the time complexity of the Chepoi-Estellon algorithm by only approximately finding diametrical pairs of vertices, that is via finding locally diametrical pairs. In the proofs of our main results, we will repeatedly apply this idea, showing that it is sufficient to solve the easier and computationally more efficient approximate version of this expensive sub-problem. We then move on to proofs of Theorems 7 and 8 in Sects. 4 and 4.1, respectively.

---

[2] The cited result also gives rise to an algorithm for general δ-hyperbolic spaces whose running time depends on the time to compute $F_S(x)$ for $x \in X$ and $S \subseteq X$. Because our interest is primarily in graphs, we direct the reader to [3] for details.

Recall from Sect. 2.1 that a pair of vertices $\{u, v\}$ is *locally diametrical* if there is no vertex w such that $d(u, w) > d(u, v)$ or $d(v, w) > d(v, u)$. Clearly a diametrical pair is locally diametrical but the converse is not true in general (e.g., a cycle with handles). It turns out to be sufficient to find locally diametrical pairs in the main lemma of [3]. Indeed, the following lemma is simply Lemma 1 from [3], but with the requirement that u and v be diametrical replaced with the weaker property of being locally diametrical.

Lemma 2. *Let X be a δ-hyperbolic metric space and $S \subseteq X$ be a compact set and $r \in \mathbb{R}$. Suppose that u and v are locally diametrical in S and let $[u, v]$ be a geodesic. Let $c = [u, v][r]$. Then $B_{2r}(u) \cap S \subseteq B_{r+\delta}(c) \cap S$.*

The proof of Lemma 1 in [3] works essentially unchanged to prove Lemma 2 by replacing diametrical pairs with locally diametrical pairs. Since we will use a refined version of the same argument that is needed for Lemma 2 in the proof of Theorem 7, we skip the proof of Lemma 2. We prove below (Lemma 4) that we can find a locally diametrical pair with at most $2\delta + 1$ breadth-first searches. Hence, we achieve the following significant reduction in the run time of the Chepoi-Estellon algorithm.

Lemma 3. *Let G be a δ-hyperbolic graph and p an integer. There exists an algorithm to find a set of p points that $(r_p + \delta)$-dominates $V(G)$ that runs in time $O(n^2 \log_2(\mathrm{diam}(G))(2\delta + 1))$.*

It remains to show how to efficiently determine locally diametrical pairs.

Lemma 4. *Given a δ-hyperbolic graph G and $S \subseteq V(G)$. There is an algorithm that finds a locally diametrical pair of vertices by performing at most $2\delta + 1$ breadth-first searches; that is, the running time is $O((2\delta + 1)(m + n))$.*

Proof. Choose a vertex $u \in S$ arbitrarily and find a vertex $v_1 \in F_S(u)$ by BFS. Then, find $v_2 \in F_S(v_1)$. Next, find a vertex $v_3 \in F_S(v_2)$. If $d(v_2, v_3) = d(v_1, v_2)$, then let $v = v_1$ and $w = v_2$ and we have found a locally diametrical pair. Otherwise $d(v_2, v_3) > d(v_1, v_2)$ and continue the process until v_k, v_{k+1} are found such that $d(v_k, v_{k+1}) = d(v_k, F_S(v_k))$ and $d(v_k, v_{k+1}) = d(v_{k+1}, F_S(v_{k+1}))$. This must happen for at most $k \leq \mathrm{diam}(S)$. But by Proposition 3 in [2] $d(v_1, v_2) \geq \mathrm{diam}(S) - 2\delta_{4-point} \geq \mathrm{diam}(S) - 2\delta$ so k cannot exceed 2δ. This means no more than $(2\delta + 1)$ BFS steps or no more than $O(2\delta + 1)(m + n)$ steps are needed for finding a locally diametrical pair starting from $u \in S$. Then algorithm returns the locally diametrical pair (v_k, v_{k+1}).

4   Approximating p-centers

In general, in searching for p-centers, first we approximately solve the dual problem, that is, we find D, a $(p + 1)$-packing, with $|D| \geq p + 1$ such that

$$\{\max \; r \mid d(s, s') \geq r, \; \forall s \neq s' \in D\} \leq d_{p+1}(V).$$

This together with Observation 4 yields

$$\frac{1}{2}\{\max \ r \mid d(s,s') \geq r, \ \forall s \neq s' \in D\} \leq r_p(V). \tag{1}$$

Given these $(p+1)$-points we find a set of p-points C such that setting $\lambda = \frac{1}{2}\{\max \ r \mid d(s,s') \geq r, \ \forall s \neq s' \in D\}$,

1. C λ-dominates the points in D, and
2. for each $a \in D$ there exists some $a' \in D$ and $c \in C$ such that c is on a geodesic between a and a'.

We prove later that these two properties together with δ-hyperbolicity allow us to show that for a carefully-selected set D, the p points in C $(\lambda + 3\delta)$-dominate V, that is,

$$\{\min \ r \mid for \ each \ x \in V, \ \exists c \in C \ with \ d(x,c) \leq r\} \leq \lambda + 3\delta. \tag{2}$$

Substituting the value of λ in (2) and applying (1) yields,

$$\{\min \ r \mid for \ each \ x \in V, \quad \exists c \in C \ with \ d(x,c) \leq r\}$$
$$\leq \frac{1}{2}\{\max \ r \mid d(s,s') \geq r, \ \forall s \neq s' \in D\} + 3\delta$$
$$\leq r_p(V) + 3\delta.$$

It follows that C $(r_p(V) + 3\delta)$-dominates V as desired.

We now apply this approach to find a 1-center of a graph.

Theorem 9. *Let G be a δ-hyperbolic graph. There exists an algorithm to find a point c that $(r_1 + \delta)$-dominates $V(G)$. The algorithm requires time $O((2\delta + 1)(m + n))$.*

Proof. Let x, y be a locally diametrical pair of vertices and let $[x, y]$ be a geodesic segment. As described above, set $\lambda = \frac{d(x,y)}{2}$ and choose $c = [x,y][\lambda]$. Clearly, $C = \{c\}$ satisfies Properties 1 and 2 above. We now show that $C = \{c\}$ $(\lambda + \delta)$-dominates V.

Let z be any point in V and consider the geodesic triangle $\Delta(x, y, z)$ as depicted and labeled in Fig. 2. Without loss of generality, assume that $d(y, z) \leq d(x, z)$. Since (x, y) is locally diametrical, then

$$d(y, z) \leq d(x, z) \leq d(x, y)$$

which implies that

$$\alpha_z \leq \alpha_y \leq \alpha_x.$$

(This means that in the figure c lies to the right of m_z, as shown.) Then

$$d(z, c) \leq \alpha_z + \delta + d(c, m_z) \leq \alpha_z + \delta + \lambda - \alpha_y \leq \delta + \lambda.$$

As the claim holds for any z, c $(\lambda + \delta)$-dominates $V(G)$, and therefore, since $\lambda = \frac{1}{2}d(x,y) \leq \frac{1}{2}d_2(V) \leq r_1(V)$, the latter inequality by Observation 4, and thus c $(r_1 + \delta)$-dominates $V(G)$, as desired. To complete the proof, we note that by Lemma 4, x and y can be found in time $O((2\delta + 1)(m + n))$.

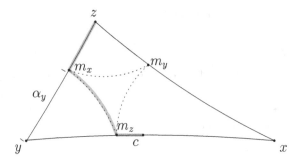

Fig. 2. A geodesic triangle $\Delta(x, y, z)$ with internal points m_x, m_y, m_z and c labelled as in the proof of Theorem 9. Dashed lines indicate a distance $\leq \delta$ and the red line indicates the upper estimate for $d(z, c)$.

We note that in the course of the above prove we demonstrated the following fact that we shall reuse.

Observation 10. *Let z be any vertex in $V(G)$, (x, y) a locally diametrical pair of vertices, $c \in A(G)$ the mid-point of $[x, y]$ and $\lambda = \frac{d(x,y)}{2}$. Then $d(z, c) \leq \lambda + \delta$.*

In extending these proof techniques to the general case for $p > 1$, we run into the following two difficulties, each costing us an additional δ in our approximation error. First, Property 2 only guarantees that p of the $\binom{p+1}{2}$ pairs of points in D have a geodesics connecting them containing some point $c_i \in C$. This will force us use two geodesic triangles to bound the distance from some points in V to their closest center in C. Second, in achieving the quasilinear runtime, we are only able to find a $(\lambda + 2\delta)$-approximation for the $(p + 1)$-packing problem. We omit further details until Sect. 4.1.

To finish off this section, we prove that when $p = 2$ we can find a 2-center solution which $(r_2 + \delta)$-dominates G. Like Theorem 9, this is stronger than our general result (Theorem 7) and the proof does not use the machinery outlined at the beginning of Sect. 4 that relies on Properties 1 and 2. Theorems 9 and 11 may be special cases of a general and stronger result than our main result, so we include it.

Theorem 11. *Let G be a δ-hyperbolic graph. There exists an algorithm to find points c_1, c_2 that $(r_2 + \delta)$-dominate $V(G)$. The algorithm requires time $O((2\delta + 1)(m + n))$.*

Proof. Let x, y be a locally diametrical pair of vertices and let $[x, y]$ be a geodesic segment. Choose z so that $\min\{d(z, x), d(z, y)\}$ is maximized (requires two BFS). We let our 3-packing be $D = \{x, y, z\}$. Assume without loss of generality that $d(x, y) \geq d(x, z) \geq d(y, z)$, and so, $\lambda = \frac{1}{2}\{\max\ r \mid d(s, s') > r,\ \forall s \neq s' \in D\} = \frac{1}{2}d(y, z)$.

We choose $c_1 = [x,y][\lambda]$ and $c_2 = [y,x][\lambda]$. We claim that $C = \{c_1, c_2\}$ satisfy Eq. 2, with $t = 1$, and so, C ($r_2 + \delta$)-dominates G.

To prove the claim, let $\Delta_1 = \Delta(x,y,z)$ be a geodesic triangle. Let w be any point of G and let $\Delta_2 = \Delta(x,y,w)$ be a geodesic triangle so that Δ_1 and Δ_2 share the geodesic $[x,y]$. We will show that $\min\{d(w,c_1), d(w,c_2)\} \le \lambda + \delta$. Take $\alpha_x, \alpha_y, \alpha_w$ and m_x, m_y, m_w to denote the internal distances and points in Δ_2. Without loss of generality assume $d(w,x) \le d(w,y)$ which implies that $d(w,x) \le d(y,z) = 2\lambda$ and $\alpha_x \le \alpha_y$. We distinguish two cases, as illustrated in Fig. 3.

Case 1: $\lambda < \alpha_x < d(x,y) - \lambda$

From the choice of z, it follows that either $d(w,x) \le d(y,z) = 2\lambda$ or $d(w,y) \le 2\lambda$. Assume without loss of generality that $d(w,x) = d(w,m_y) + d(m_w,x) \le 2\lambda$. Therefore, $d(w,c_1) \le d(w,m_y) + d(m_y,m_w) + d(m_w,c_1) \le d(w,m_y) + \delta + d(m_w,x) - \lambda \le \lambda + \delta$.

Case 2: $\alpha_x \le \lambda$

In this case m_w lies between x and c_1 on the geodesic segment $[x,y]$. By the local maximality of x and y, we have $d(y,w) = \alpha_y + \alpha_w \le \alpha_y + \alpha_x = d(x,y)$ and so $d(w,m_y) = \alpha_w \le \alpha_x = d(x,m_w)$. Then $d(w,c_1) \le d(w,m_y) + d(m_y,m_w) + d(m_w,c_1) \le d(x,c_1) + \delta = \lambda + \delta$.

To complete the proof, we need only show that c_1, c_2 can be found in $O((2\delta + 1)(m+n))$ time. By Lemma 4, x and y can be found in time $O((2\delta + 1)(m+n))$ and the vertex z can be found by doing a breadth-first search rooted at x and one rooted at y. Given $D = \{x,y,z\}$, the vertices c_1 and c_2 can then be found by storing the last breadth-first search used in finding x and y and $\lambda = \frac{1}{2}\min\{d(x,x), d(y,z)\}$. The runtime now follows.

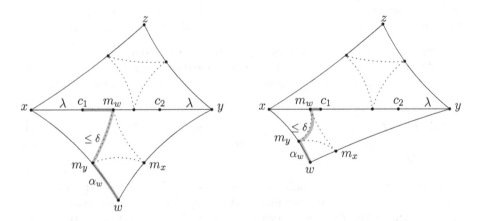

Fig. 3. Figure for Cases 1 and 2 in the proof of Theorem 11. The red lines indicate the upper estimate for $d(w,c_1)$. Dashed lines indicate a distance $\le \delta$.

4.1   The General Algorithm

Our algorithm and proof follow the same three basic steps, though each step is more involved. As a reminder these three steps are (1) approximately solving the dual problem, or finding a $(p + 1)$-packing, (2) deriving p-points from this dual solution that satisfy Properties 1 and 2, and (3) bounding the approximation guarantee by showing Eq. 2.

It turns out the difficult part of these three steps is Step 1. For this step, we need to extend the notion of a 'locally diametrical pair' to a 'locally diametrical set' in such a way that i) it provides us with both the tools we need to satisfy Properties 1 and 2 and ii) it can be determined efficiently. We find a set of $(p+1)$ vertices $D = \{v_0, v_1, \ldots, v_p\}$ with

$$\lambda(D) := \tfrac{1}{2}\{\max\ r \mid d(s, s') \geq r,\ \forall v_i \neq v_j \in D\}$$

such that the following three properties hold

(a) (Vertex relabeling) $d(v_0, v_i) = 2\lambda(D)$ for some $v_i \in D$,
(b) (Extending locally diametrical pairs to *locally diametrical sets*) For each $v_i \in D$ with $d(v_i, v_j) = 2\lambda(D)$ for some v_j, there exists no $w \in V(G)$ with $d(w, v_k) > 2\lambda(D), \forall v_k \in D \setminus \{v_i\}$, and
(c) (δ-hyperbolic version of locally diametrical sets) for each $i \geq 1$, there exists no vertex $v \in V(G)$ with $d(v_0, v) > d(v_0, v_i) + 2\delta$ and $d(v_i, v) \leq 2\lambda(D)$ and $d(v, v_j) > 2\lambda(D)$ for each $j \neq i$.

These three requirements provide us with what is needed to determine a set of $(p + 1)$ vertices satisfying Properties 1 and 2. Specifically, we prove

Lemma 5. *Let G be a δ-hyperbolic graph and $\Lambda_n = \lceil 4 + 3\delta + 2\delta \log_2 n \rceil$. There exists an algorithm to find a set D of $p + 1$ vertices satisfying (a), (b) and (c). The algorithm runs in time $O(n \log_2 n + (m + n)((2p + 1)\Lambda_n + (p + 1)))$.*

Given a set of $p+1$ vertices satisfying Properties (a), (b) and (c) it is straightforward to find $C = \{c_1, \ldots, c_p\}$ satisfying Properties 1 and 2. For each $1 \leq i \leq p$, let c_i be the vertex at distance λ from v_i on the shortest path from v_i to v_0, i.e. $c_i = [v_i, v_0][\lambda]$.

Lemma 6. *Let G be a δ-hyperbolic graph. Suppose that $D = \{v_0, v_1, \ldots, v_p\}$ satisfy (a), (b) and (c). Then the set of p points $C = \{c_i \mid c_i = [v_i, v_0][\lambda]\}$ $(\lambda + 3\delta)$-dominate G.*

As described above (beginning of Sect. 4), such C $(r_p(V) + 3\delta)$-dominates V as desired. So, given the Lemmas 5 and 6, the proof of Theorem 7 follows once establishing the runtime, which we do now. First, determining the set D takes $O(n \log_2 n + (m + n)((2p + 1)\Lambda_n + (p + 1)))$. Given D, the set of vertices $\{c_i, 1 \leq i \leq p\}$ can clearly be constructed by performing a breadth-first search rooted at v_0. Theorem 7 now follows.

It remains to establish Lemmas 5 and 6. Lemma 5 is the more interesting of the two proofs, and takes us deeper into the analysis of locally diametrical sets. The proof of Lemma 6 is a sophistication of the ideas in Theorems 9 and 11. The proof of that lemma appears in [5] Sect. 4.2.

A complete proof of Lemma 5 is found in [5] Sect. 4.3, but we now give a quick outline of its ideas. The first step is to show that we can a find $(p+1)$-packing that is within $O(\delta \log_2 n)$ of an optimal solution. To find this initial $(p+1)$-packing, we deduce from previously known results that in time $O(n \log_2 n)$ we can find a set $\mathcal{P}$ of $p+1$ vertices that is a κ-dispersion, for some $\kappa \geq d_{p+1}(G) - \Lambda_n$. To do so, we find a tree T which approximately preserves distances on our input graph G. It turns out that exactly solving the $(p+1)$-packings on trees can be done efficiently, though in contrast to before, we solve the p-centers first and use this to construct a dual solution in G. The fact that T is a good approximating tree allows us to bound how close our $(p+1)$-packing is to an optimal solution and in turn helps us achieve the quasilinear running time.

Then, given this initial $(p+1)$-packing $\mathcal{P}$, we iteratively improve the solution whenever possible until we achieve Properties (a), (b), (c). Clearly, (a) can hold for all solutions after relabelling, so the only difficulty is in ensuring both (b) and (c) hold. In fact, it is not too hard to make either one of (b) or (c) hold by making local improvements: simply scan the vertices in our set and if we find a vertex witnessing that our desired property is false, we replace it. Using the 4-point condition (Lemma 1), we show that with a bounded number of replacements we obtain a set satisfying the property we are interested in. To satisfy (b) and (c) simultaneously, we alternate applications of subroutines to satisfy (b) and (c) until both properties hold. We use a potential function argument to show that this process terminates after a number of replacements bounded by $O(p\delta \log_2 n)$, and our running time follows.

5   Experimental Results

To demonstrate scalability, we have implemented the algorithms from Theorem 7 ($p \geq 3$) and Theorem 8 ($p \leq 2$). For comparison, we have also implemented the algorithms of Chepoi et al. (abbreviated Ch.) [2] ($p = 1$) and Chepoi-Estellon (abbreviated C-E) [3] ($p \geq 2$). We also compared Theorems 7 and 8 to the following simple algorithm: Compute a distance approximating tree T as described in Sect. 4 of [2] and return an exact solution to p-centres on T (such a tree has vertex set $V(G)$ and approximates pairwise distances in G to within $O(\delta \log_2 n)$).

We ran the algorithms on four graphs extracted from real networks arising from different types of data. All graphs are simple and have unit edge lengths and each has a small hyperbolicity constant. Table 1 briefly summarizes the networks we analyzed; more information about the data can be found in [13].[3] In the case

---

[3] The graphs p2p-gnutella25 and web-stanford are available publicly as part of the Stanford Large Network Dataset Collection. The sn-medium graph is extracted from the social network Facebook, and the sprintlink-1239 graph is an IP-layer network from the Rocketfuel ISP.

Table 1. Networks analyzed

Network	Type	$|V|$	$|E|$	Diameter	Radius	$\delta_{4-point}$
sprintlink-1239	Rocketfuel ISP network	8341	14025	13	7	3
p2p-gnutella25	Peer-to-peer network	22663	54693	11	7	3
sn-medium	Social network	26567	226566	14	7	4
web-stanford	Web network	255265	1941926	164	82	1.5 (est.)

Table 2. Comparison of estimates of the p-radius.

	sprintlink-1239			p2p-gnutella25			sn-medium			web-stanford		
	Thm 9	Ch.	Tree	Thm 9	Ch.	Tree	Thm 9	Ch.	Tree	Thm 9	Ch.	Tree
$p=1$	7	7	8	8	8	7	7	7	8	82		
$p=2$	7	7	7	8	8	7	7	8	8	59		
	Thm 8	C-E	Tree	Thm 8	C-E	Tree	Thm 8	C-E	Tree	Thm 8	C-E	Tree
$p=3$	5	6	6	7	7	7	7	7	8	47		
$p=4$	5	6	6	7	7	7	6	7	8	46		
$p=5$	4	5	6	7	6	7	6	7	8	44		
$p=6$	4	5	6	7	6	7	6	6	8	44		
$p=7$	4	5	5	6	6	7	6	6	8	44		
$p=8$	4	5	5	6	6	7	6	6	8	38		
$p=9$	4	5	5	6	6	7	6	6	7	29		
$p=10$	4	5	5	6	6	7	6	6	7	29		
$p=15$	4	4	5	6	5	7	5	6	7	22		
$p=20$	4	4	5	5	5	7	5	6	7	17		

of the web-stanford graph, we have only an estimate of $\delta_{4-point}$ obtained by sampling since the graph is quite large. Table 2 contains a comparison of the estimated p-radius r_p of the three algorithms. We have run only our algorithm from Sect. 4.1 on the largest network (web-stanford), since the running time of Chepoi-Estellon is infeasible on a graph of this size.

Our experiments indicate that, as far as accuracy goes, our algorithm performs similarly to that of Chepoi-Estellon despite the larger theoretical upper bound on the error. In many cases our estimate is actually better than that one. The p-radius estimated by Sect. 4.1 is always within 1 of their estimate in our trials. Combined with the significant improvement in running time, this makes our algorithm a preferable choice for solving p-centres in practice. For comparison, our implementation of our algorithm terminated in under two seconds on the sn-medium graph, while Chepoi-Estellon took about one minute.

While the tree-approximation heuristic is simple, and runs in quasilinear time $O(m + n)$, the approximation guarantee is only as good as the distance approximation of T, hence the additive error could up to $O(\delta \log n)$. However, our experiments show that it seems to perform well in practice and may be a good choice of heuristic in some applications.

Acknowledgement. The authors are grateful to the anonymous reviewers for helpful comments.

References

1. Bridson, M.R., Haefliger, A.: Metric Spaces of Non-positive Curvature, vol. 319. Springer Science & Business Media, Berlin (1999)
2. Chepoi, V., Dragan, F., Estellon, B., Habib, M., Vaxès, Y.: Diameters, centers, and approximating trees of delta-hyperbolic geodesic spaces and graphs. In: Proceedings of 24th Annual Symposium on Computational Geometry, pp. 59–68. ACM (2008)
3. Chepoi, V., Estellon, B.: Packing and covering δ-hyperbolic spaces by balls. In: Charikar, M., Jansen, K., Reingold, O., Rolim, J.D.P. (eds.) APPROX and RANDOM 2007. LNCS, vol. 4627, pp. 59–73. Springer, Heidelberg (2007)
4. Dyer, M.E., Frieze, A.M.: A simple heuristic for the p-centre problem. Oper. Res. Lett. **3**(6), 285–288 (1985)
5. Edwards, K., Kennedy, W.S., Saniee, I.: Fast approximation algorithms for p-centres in large δ-hyperbolic graphs (2016). arXiv preprint arXiv:1604.07359
6. Erkut, E.: The discrete p-dispersion problem. Eur. J. Oper. Res. **46**(1), 48–60 (1990)
7. Erkut, E., Neuman, S.: Comparison of four models for dispersing facilities. INFOR **29**, 68–86 (1991)
8. Erkut, E., Ülküsal, Y., Yenicerioglu, O.: A comparison of p-dispersion heuristics. Comput. Oper. Res. **21**(10), 1103–1113 (1994)
9. Gromov, M.: Hyperbolic Groups. Springer, Berlin (1987)
10. Hochbaum, D.S., Shmoys, D.B.: A best possible heuristic for the k-center problem. Math. Oper. Res. **10**(2), 180–184 (1985)
11. Hsu, W.-L., Nemhauser, G.L.: Easy and hard bottleneck location problems. Discrete Appl. Math. **1**(3), 209–215 (1979)
12. Johnson, D.B.: Efficient algorithms for shortest paths in sparse networks. J. ACM (JACM) **24**(1), 1–13 (1977)
13. Kennedy, W.S., Narayan, O., Saniee, I.: On the hyperbolicity of large-scale networks. arXiv e-prints, June 2013
14. Ravi, S.S., Rosenkrantz, D.J., Tayi, G.K.: Facility dispersion problems: heuristics and special cases. In: Dehne, F., Sack, J.-R., Santoro, N. (eds.) WADS 1991. LNCS, vol. 519, pp. 355–366. Springer, Heidelberg (1991). doi:10.1007/BFb0028275
15. Shier, D.R.: A min-max theorem for p-center problems on a tree. Transp. Sci. **11**(3), 243–252 (1977)

Graphon-Inspired Analysis on the Fluctuation of the Chinese Stock Market

Linyuan Lu[1,2(✉)], Arthur L.B. Yang[2], and James J.Y. Zhao[3]

[1] Department of Mathematics, University of South Carolina,
Columbia, SC 29208, USA
lu@math.sc.edu
[2] Center for Combinatorics, LPMC-TJKLC,
Nankai University, Tianjin 300071, People's Republic of China
yang@nankai.edu.cn
[3] Center for Applied Mathematics,
Tianjin University, Tianjin 300072, People's Republic of China
jjyzhao@tju.edu.cn

Abstract. The Chinese A-share Stock Market has been suffering from massive volatility since the popping of a bubble on June 15, 2015. About a third of the values of A-shares in Shanghai Stock Exchange was lost within one month of the event. Although the Chinese government enacted many measures to halt the fall, the turbulence of the Chinese Stock Market continues in 2016. Motivated by the theory of graph limits, we apply motif statistics, dual motif statistics, and cut distances to study the correlation network structures of the Chinese A-share Stock Market in the last two years. The changing patterns of our measures match the major events of Chinese stock market. Our method extends the traditional motif-based method.

Keywords: Chinese A-share Market · Correlation network · Motif statistics · Dual motif statistics · Cut distance

1 Introduction

Chinese A-shares are shares of the Renminbi currency that are purchased and traded on the Shanghai and Shenzhen stock exchanges. A-shares are mainly owned by Chinese citizens. Since China is now such a big force in the global economy, the turmoil that the Chinese A-share Stock Market had experienced in the past year has inevitably affected the rest of the world. The turbulence of the Chinese A-share Stock Market began with the popping of a bubble on June 15, 2015. About a third of the values of the A-shares on the Shanghai Stock Exchange was lost within one month of the event. Although the Chinese government enacted many measures to halt the crash of the stock market, the turbulence continued to 2016. In January 2016 the Chinese A-share stock market

L. Lu—This author was supported in part by NSF grant DMS 1600811.

experienced a steep sell-off and trading was halted on January 4 and 7 due to the new adopted circuit breaker mechanism after the market fell 7 % within the first 30 minutes. We collect the A-share data from the Shanghai Stock Exchange (SSE) and the Shenzhen Stock Exchange (SZSE) using the Wind Financial Terminal during the period from December 30, 2013 to April 29, 2016. The weekly distributions of SSE Composite Index and SZSE Component Index are shown in Fig. 1. The peak point occurs during the week of June 8–12, 2015, and is followed by a sudden sharp drop in next 3 weeks. Another sharp drop occurs in January 2016. (See Sect. 4 for further discussion.)

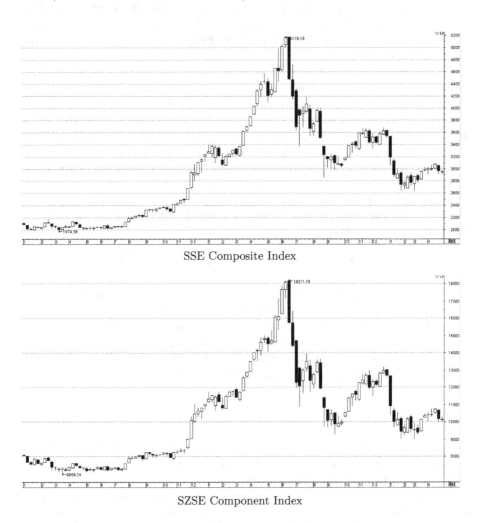

SSE Composite Index

SZSE Component Index

Fig. 1. The weekly distributions of SSE composite index and SZSE compoent index from December 30, 2013 to April 29, 2016.

To better understand the fluctuation of the Chinese A-share Stock Market, we will construct a sequence of correlation graphs out of raw stock data. Then we will use graphon-inspired graph parameters, including classical subgraph/motif counts, to analyze these graphs. Let us review relevant methods in the literature.

Network based analyses are common to study the relationships between financial entities. Mantegna [14] first applied network-based analysis on the DJIA stocks (Dow Jones Industrial Average index) and the S&P-500 stocks (Standard and Poor's 500 index). Onnela, Kaski and Kertész [17] constructed the asset graph on NYSE stocks (New York stock exchange). In this paper, we use the correlation threshold method developed in [5,11,16]. Huang, Zhuang and Yao [11] studied many graph properties of the correlation graph about the Chinese Stock Market, including degree distribution, cluster coefficients, connected components, cliques, etc.

Subgraph counting (also known as network motif statistics) are often used to study the structure and function of biological and online social networks. In 2002, Milo et al. [15] defined *network motifs* to be patterns of inter-connections occurring in complex networks at numbers that are significantly higher than those in randomized networks. Sporns and Kötter [19] studied motifs in Brain Networks. Alon [1] studied motifs for various biological networks, including signaling and neuronal networks. Juszczyszyn et al. [12] used motifs to study social networks. Onnela et al. [18] studied motifs in weighted complex networks and applied them to financial and metabolic networks. Efficient sampling methods used for estimating motif statistics are studied by Bhuiyan et al. [3], Birmelé [4], and Wang et al. [20], etc.

The theory of graph limits (i.e. graphons) have been developed by Borgs, Chayes, Lovász, Sós, Vesztergombi, and others, rapidly in the last decade. It has many important theoretical applications including quasi-random graphs, Turán density, statistical physics, etc. Motivated by the theory of graph limits, we suggest a set of additional graph parameters so called "dual motif statistics" to study the stock market. The paper is organized as follows: In Sect. 2, we review the background of graph limits. In Sect. 3, we give some theoretical results on the efficiency of simple sampling algorithm. Finally, in Sect. 4, we compute several motif and dual motif statistics for the correlation stock networks.

2   Graph Limits

To understand how graphs fluctuate, it is necessary to know when a graph sequence converges. This is exactly the scope of the theory of graph limits. Here, we only focus on the dense graphs.

A graph $G = (V, E)$ consists of a vertex set V and an edge set E (Loops are allowed here). The number of vertices of G is denoted by $v(G)$ (or by $|G|$) while the number of edges of G is denoted by $e(G)$. Given two graphs H and G, a *graph homomorphism* from H into G is a map $f: V(H) \to V(G)$ keeping the edges, i.e., $f(u)f(v) \in E(G)$ whenever $uv \in E(H)$. Let $hom(H, G)$ be the number of homomorphisms from H into G, $inj(H, G)$ the number of *injective* homomorphisms of H into G, and $ind(H,G)$ the number of embeddings of H into G as induced subgraphs.

The homomorphism density is defined by

$$t(G,H) = \frac{hom(G,H)}{v(H)^{v(G)}},$$

which is the probability that a random map f from $V(G)$ to $V(H)$ is a graph homomorphism. Assuming $v(G) = n > k = v(H)$, the *subgraph density* is defined by

$$t_{inj}(H,G) = \frac{inj(H,G)}{(n)_k}.$$

Here $(n)_k = n(n-1)\ldots(n-k+1)$ is the k-th falling power of n. Similarly, the *induced subgraph density* is defined by

$$t_{ind}(H,G) = \frac{ind(H,G)}{(n)_k}.$$

Let $\{G_n\}$ be a sequence of dense graphs. The following three definitions of the convergence of $\{G_n\}$ were proved to be equivalent by Borgs-Chayes-Lovász-Sós-Vesztergombi [6–8] (also see Lovász's book [13]).

Convergence from Left: For any fixed graph F, $\lim_{n\to\infty} t(F,G_n)$ exists.
Convergence from Right: For any fixed graph F, $\lim_{n\to\infty} t(G_n,F)$ exists.
Convergence under cut-distance: Under the cut-distance norm (defined later), $\{G_n\}$ form a Cauchy sequence.

Note that in the first item "Convergence from Left", one can also replace $t(F,G_n)$ by $t_{inj}(F,G_n)$ or $t_{ind}(F,G_n)$. The connections among $t(F,G_n)$, $t_{inj}(F,G_n)$, and $t_{ind}(F,G_n)$ are described in the following equations:

$$t(F,G) \approx t_{inj}(F,G) \quad \text{when } |F| \text{ is small and } G \text{ is large;}$$
$$t_{inj}(F,G) = \sum_{F'\subseteq F} t_{ind}(F',G).$$

The limit objects are called graphons, which are symmetric measurable functions from $[0,1]^2$ to $[0,1]$ modulo certain equivalence relationship. Graphons can be visually described as pixel maps on the unit square. A pixel map of a graph G is a visual representation of its adjacency matrix where 1 is represent by a black square and 0 is represented by a white square. For example, Fig. 2 is the pixel map of the 5-cycle C_5.

The classical Erdős-Renyi random graph model, $G(n,p)$, is the random graph G on n vertices so that each pair of vertices is an edge of G with probability p independently. For a fixed constant p, the limit object of $\{G(n,p)\}$, denoted by $G(\infty,p)$, is a constant function which takes value p over all the unit square.

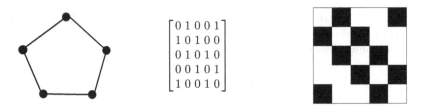

Fig. 2. Graph C_5, its adjacency matrix, and the pixel map.

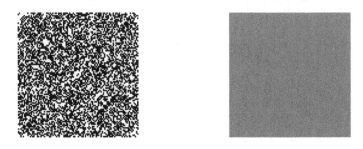

Fig. 3. The pixel maps of a random graph $G(100, 0.5)$ and the graphon $G(\infty, 0.5)$.

In Fig. 3, here are the pixel maps of $G(100, 0.5)$ and the graphon $G(\infty, 0.5)$ respectively.

Let A be an $n \times n$ matrix. The *cut norm* of A, introduced by Frieze and Kannan [10], is defined by

$$\|A\|_\square = \frac{1}{n^2} \max_{S,T \subseteq [n]} \left| \sum_{i \in S, j \in T} A_{ij} \right|.$$

For two graphs G and G' with a common vertex set $[n]$, their cut distance is defined by

$$d_\square(G, G') = \|A_G - A_{G'}\|_\square,$$

where A_G and $A_{G'}$ are the adjacency matrix of G and the adjacency matrix of G' respectively. The definition of the cut distance between two graphs with different set of vertices (or a different number of vertices) is irrelevant to this paper and will be omitted here. For interested readers, please read Chap. 8 of [13].

3   Efficient Sampling Algorithm

Observe that computing $t_{ind}(F, G)$ is the same as subgraph counting (or motif statistics). These parameters have been widely used for various complex graphs [1,3,4,12,15,19,20]. Here we suggest another set of parameters $t(G, F)$, where G

is the target graph and F is a small fixed graph. Since the theory of graph limits applies to sequences of graphs of increasing size, we cannot say that two family of parameters $\{t(G, F)\colon F \text{ is small}\}$ and $\{t_{ind}(F, G)\colon F \text{ is small}\}$ are equivalent. Nevertheless, $t(G, F)$ seems to be another interesting graph parameters which we refer to as *dual motif statistics*.

By definition, the exact value of $t(G, F)$ is the average of $|F|^{|G|}$ 0–1 terms. Under the assumption that $|G|$ is large and $|F|$ is small, this is difficult to compute the exact value. In comparison, computing $t_{ind}(F, G)$ only requires at most $|G|^{|F|}$ steps, which is polynomial time in $|G|$ given a fixed size $|F|$. So sampling algorithm is needed here.

The *law of large numbers* states for the case where X_1, X_2, ..., is an infinite sequence of i.i.d. Lebesgue integrable random variables with expected value $E(X_1) = E(X_2) = \cdots = \mu$, then the sample average

$$\bar{X}_n = \frac{1}{n}(X_1 + \cdots + X_n)$$

converges to the expected value

$$\bar{X}_n \to \mu \text{ for } n \to \infty.$$

To apply the law of large numbers, let X be an indicator random variable that a random mapping $f\colon V(G) \to V(F)$ is a graph homomorphism. Then the expectation of X is simply $E(X) = t(G, F)$. The following result states how many times of sampling is needed for required accuracy.

Theorem 1. *For any $\epsilon > 0$, for any 0–1 random variable X, with probability at least $1 - \epsilon$, the sample average $\bar{X}_n$ satisfies*

$$|\bar{X}_n - E(X)| \leq \sqrt{\frac{2 \log(2/\epsilon)}{n}}.$$

Proof. We use the following version of Chernoff's inequality [9]. Let $S_n = X_1 + \cdots + X_n$ be the sum of independent non-negative random variables. Then,

$$\Pr(S_n - E(S_n) < -\lambda) < e^{-\frac{\lambda^2}{2E(S_n)}}.$$

Choose $\lambda = \sqrt{2n \log(2/\epsilon)}$ and note that $E(S_n) = nE(X) \leq n$. We conclude with

$$\Pr(S_n - E(S_n) < -\lambda) < \frac{\epsilon}{2}.$$

Now we apply the same argument to $n - S_n = \sum_{i=1}^{n}(1 - X_i)$, which is still the sum of n independent non-negative random variables. We get

$$\Pr(S_n - E(S_n) > \lambda) = \Pr((n - S_n) - E(n - S_n) < -\lambda) < \frac{\epsilon}{2}.$$

Thus, with probability at least $1 - \epsilon$, we have

$$|S_n - E(S_n)| \leq \lambda.$$

Equivalently, with probability at least $1 - \epsilon$, we have

$$|\bar{X}_n - t(G, F)| \leq \frac{\lambda}{n} = \sqrt{\frac{2 \log(2/\epsilon)}{n}}.$$

Remark: When applying this theorem to estimate $t(G, F)$, we observe that the number of sampling required only depends on the accuracy, and is independent of the sizes of $|G|$ and $|F|$. Thus, $t(G, F)$ can be estimated efficiently, just as the motif statistics $t_{ind}(F, G)$.

4   Analysis on Correlation Networks of the Chinese A-Share Stocks

Here we briefly review the correlation threshold method to construct the stock correlation graph. Let $P_i(t)$ be the price of stock i at time t. Then the return of the price at a time interval $(t - \Delta t, t)$ is defined as

$$r_i(t) = \ln P_i(t) - \ln P_i(t - \Delta t).$$

We will take Δt as one day so that $r_i(t) = \ln(1 + p_i)$ where p_i is the percentage of the price changing of the i-th stock.

For a given time interval $[a, b]$, let r_i be the vector so that the index t runs from a to b. We view r_i as the incidences of a random variable, which is also denoted by r_i. The cross-correlations between the i-th stock and the j-th stock are given by:

$$c_{ij} = \frac{\text{coVar}(r_i, r_j)}{\sqrt{\text{Var}(r_i)\text{Var}(r_j)}}. \tag{1}$$

Then c_{ij} can vary between $[-1, 1]$, where 1 (or -1) means that two stocks i and j are completely correlated (anti-correlated). To construct a graph from this matrix, one may use the threshold method. For a given threshold $\tau < 1$, adding edge ij to a graph G if $|c_{ij}| \geq \tau$. This graph is called the correlation stock network at time $[a, b]$ with the threshold τ.

In other literatures the time interval was chosen rather long (for example, several years), so they got long-term statistical properties of the stock market. Unlike those papers, we will focus on the fluctuation of the market rather than the long-term statistical properties, which might be more valuable for some short term investors. We choose the time interval to be one week so we get one correlation stock network per week.

We use the data of A-shares collected from SSE and SZSE for the period from December 30, 2013 to April 29, 2016. In total, there are 2832 A-share stocks in SSE and SZSE. The return for each stock is computed daily. To generate the correlation network, we set the time interval to be one week. Normally, each vector r_i has 5 entries corresponding to 5 trading days. Some weeks have significantly less trading days due to the Chinese holidays. To be consistent, we simply do not use the data in those weeks: 2/3/2014–2/7/2014, 4/28/2014–5/2/2014, 9/29/2014–10/3/2014, 10/6/2014–10/10/2014, 12/29/2014–1/2/2015, 2/16/2015–2/20/2015, 2/23/2015–2/27/2015, 8/31/2015–9/4/2015, 9/28/2015–10/2/2015, 10/5/2015–10/9/2015, and 2/8/2016–2/12/2016.

It should be noticed that the CSI 300 index (China Securities Index 300) has slumped 7% from the previous day and has triggered the circuit breaker mechanism to halt the market twice in 1/4/2016 and 1/7/2016 respectively. So the data in the week 1/4/2016–1/8/2016 was also excluded.

The threshold we used in this paper is $\tau = 0.9$. If stock i has no transaction during a week, then it has 0 correlation coefficient to other stocks.

There are 122 weeks from December 30, 2013 to April 29, 2016. After discarding the 12 weeks containing holidays and the circuit breaker, there are 110 weeks left (which are listed in Table 1 in Appendix), and we get 110 correlation stock networks $G_1, G_2, \ldots, G_{110}$. Each graph G_i has 2832 vertices. The distribution of the number of edges in G_i is shown in Fig. 4.

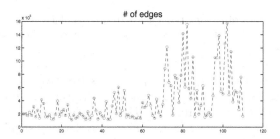

Fig. 4. The distribution of edges in the correlation networks of Chinese A-share Stock Market.

4.1 Some Motif Statistics

To illustrate how these graphs $\{G_i\}$ differ from each other, we first compute $t_{ind}(F, G_i)$ for all small graphs F over three vertices up to graph isomorphism. There are 4 of them: F_0, F_1, F_2, and F_3 (shown in Fig. 5).

Fig. 5. Four non-isomorphic graphs on three vertices.

Here are the distributions of $t_{ind}(F, G_i)$ for various F listed in items 1–4. (See Fig. 6.)

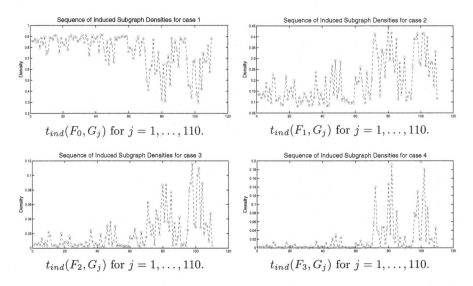

Fig. 6. The distributions of $t_{ind}(F_i, G_j)$ for $i = 0, 1, 2, 3$ and $j = 1, 2, \ldots, 110$. Here the vertical axis stands for the value of density $t_{ind}(F_i, G_j)$ and the horizontal axis stands for the graph index j.

From Fig. 6, we observe, $t_{ind}(F_0, G_j)$ reaches (relatively large) local minimum at $j = 80, 82, 98, 102$; $t_{ind}(F_1, G_j)$ reaches (relatively small) local maximum at $j = 72, 76, 80, 82, 85, 98, 101, 102$; $t_{ind}(F_2, G_j)$ (relatively small) reaches local maximum at $j = 80, 82, 85, 98, 102$; $t_{ind}(F_3, G_j)$ (relatively small) reaches local maximum at $j = 72, 82, 98, 101, 102$. We will discuss the stock market near these weeks later.

4.2   Some Dual Motif Statistics

Now we will study $t(G, F)$ for some small graph F. Since G is quite dense, almost all values of $t(G, F)$ are trivial (either 0 or 1) unless $|F|$ is large enough. To ensure non-trivial values, we consider the following variant of dual motif statistics: Let F be a small graph with loops on s vertices together with a partition of integer $|G| = n_1 + \cdots + n_s$. Consider a random partition π of $V(G) = V_1 \cup \cdots \cup V_s$ such that $|V_i| = n_i$ for $1 \leq i \leq s$. Now we associate to π a map $f \colon V(G) \to V(F)$ so that $f^{-1}(i) = V_i$. Let X be the indicator random variable so this map is a graph homomorphism. Finally, we define $t'(G, F) = \mathrm{E}(X)$. Note that Theorem 1 on sampling still works for $t'(G, F)$. We consider three such dual motifs F_4, F_5, and F_6 as specified in Fig. 7.

In Fig. 8, $t'(G_j, F_4)$ takes values very close to zero at $j = 48, 51, 71, 72, 73, 76, 77, 79, 80, 82, 85, 90, 96, 97, 98, 101, 102, 104, 109$; $t'(G_j, F_5)$ takes values very close to zero at $j = 71, 72, 73, 76, 79, 80, 82, 85, 96, 97, 98, 101, 102, 104, 109$; $t'(G_j, F_6)$ takes values very close to zero at $j = 71, 72, 80, 82, 85, 90, 96, 97, 98, 99, 100, 101, 102, 104, 106, 109$. Note that the value close to zero suggests the big fluctuation of the market in the same period.

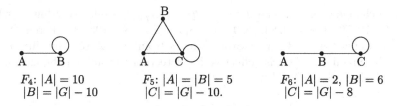

F_4: $|A| = 10$ F_5: $|A| = |B| = 5$ F_6: $|A| = 2, |B| = 6$
$|B| = |G| - 10$ $|C| = |G| - 10$. $|C| = |G| - 8$

Fig. 7. Three dual motifs F_4, F_5, and F_6.

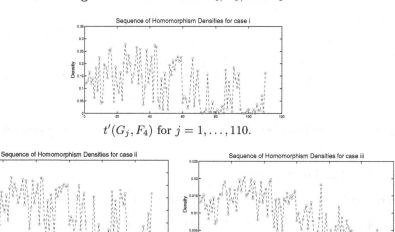

$t'(G_j, F_4)$ for $j = 1, \ldots, 110$.

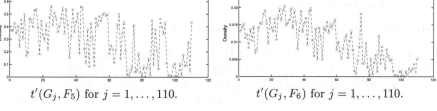

$t'(G_j, F_5)$ for $j = 1, \ldots, 110$. $t'(G_j, F_6)$ for $j = 1, \ldots, 110$.

Fig. 8. The distributions of $t'(G_j, F_i)$ for dual motif F_4, F_5, F_6 and $j = 1, 2, \ldots, 110$. Here the vertical axis stands for the density $t'(G_j, F_i)$ and the horizontal axis stands for the graph index j.

4.3 Cut Distances

Another way to capture the weekly changes of these correlation stock networks is using cut-distances. Computing the cut-distance between two graphs is NP-hard but can be approximated by a polynomial algorithm. More precisely, Alon and Naor [2] proved that the problem of appoximating the cut-norm of a given real matrix is MAX SNP hard and they presented an efficient approximation algorithm with approximation ratio $\rho > 0.56$. Their algorithm uses semi-definite programming and Grothendieck's inequality, but takes still non-trivial computing time for our purpose. Instead, we use a very simple heuristic algorithm to approximate the cut distance: first we randomly sample a partition of the vertex set; then run a greedy algorithm until a local maximum is found. We repeated these processes many times. Heuristically we found that the computed values converge very quickly. The drawback is that we do not have any theoretical results to guarantee the approximate ration like Alon-Naor's algorithm does. The approximated cut-distance between two consecutive graphs are shown in Fig. 9. The bigger the distance between the two consecutive graphs, the more change there is. The distance $d_\square(G_j, G_{j+1})$ is relatively bigger when $j = 70, 71, 72, 79, 80, 81, 82, 84, 85, 95, 97, 98, 100, 101, 102, 103, 104$. It reflects the dramatic changes on the Chinese Stock Market from June 2015. Thus in roughly saying, the cut-distance is also effective for detecting the sudden changes of the Chinese A-share Stock Market.

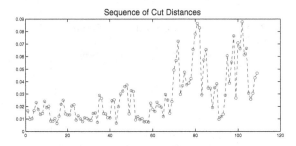

Fig. 9. The cut distance between two stock correlation networks in consecutive weeks.

4.4   Further Discussion

We observe four obvious big changes as shown in Figs. 6, 8 and 9.

1. From $t_{ind}(F, G_{46})$ to $t_{ind}(F, G_{51})$ (December 8, 2014 to January 23, 2015).
2. From $t_{ind}(F, G_{71})$ to $t_{ind}(F, G_{73})$ (June 22, 2015 to July 10, 2015).
3. From $t_{ind}(F, G_{80})$ to $t_{ind}(F, G_{85})$ (August 24, 2015 to October 23, 2015).
4. From $t_{ind}(F, G_{98})$ to $t_{ind}(F, G_{104})$ (January 25, 2016 to March 18, 2016).

There are two reasons that caused the changes in the first window. First, there are only three trading days from December 29, 2014 to January 2, 2015, since January 1 and 2, 2015 are holidays in China. Hence, that week is excluded as described in the beginning of this section. By the historical experiences, the benchmark indexes such as the SSE Composite Index or the SZSE Component Index usually fluctuate around some 'big' holidays, especially New Year's Day. Second, both indexes trend to be horizontal in the last three weeks (January 5, 2015 – January 23, 2015) after a steady increasing in previous weeks (see Fig. 1), which triggers the downfall in the following week. So the parameters capture the important moments in the stock market.

During the second window of big changes, the benchmark SSE Composite Index kept slumping in the first two weeks, but gained 5.18 % in the last week and closed at 3877.80 points on July 10, 2015 (See Fig. 1). From June 15, 2015, both the SSE Composite Index and the SZSE Component Index started plummeting, and many investors' confidences had been deeply beaten. Most of the investors sold their shares at an even lower price than their cost, which made the benchmark indexes drop even faster. In June 26, a rare thing happened that more than two thousand shares' prices hit the down-limit (−10 %). In fact, the SSE Composite Index dropped 23.37 % from June 23, 2015 (4576.49 points) to July 8, 2015 (3507.19 points). Meanwhile, the Chinese central government and the CSRC (China Securities Regulatory Commission) strained almost every nerve to rescue the stock market. Many regulation clauses and positive policies aiming to take the market back to the bull phase were released. For instance, short selling through future trading were restricted. Although it did not seem to work at first, the market slowly recovered from July 9. That day, the Ministry of Public Security of the People's Republic of China joined the rescue project. They stationed in the CSRC to investigate and deal with illegal transactions. The SSE Composite Index gained 5.76 % and closed at 3709.33 points in July 9, 2015 and kept going up until July 10, 2015.

There are two reasons in the third window. First, in 2015, there are only three trading days from August 31 to September 4, only three from September 28 to October 2, and only two from October 5 to 9. Since, September 3 is the Anti-Japanese War Victory Day of China and October 1 is the National Day of China, the whole week was a holiday. The data from three weeks mentioned above are excluded from graphs. Second, after slumping from June 15, the SSE Composite Index reached the valley of 2015 in August 26 (2850.71 points) and recovered. The SZSE Component Index also rebounded after hitting the rock bottom (9259.65 points) on September 15. In September, both the SSE Composite Index and the SZSE Component Index almost went flat. The SSE Composite Index gained 15.66 % from October 8 (3143.36 points) to November 26 (3635.55 points). The SZSE Component Index gained 22.83 % from October 8 (10394.73 points) to November 26 (12767.51 points).

For the last window, there are two causes. First, in the first week of 2016, the circuit breaker mechanism was triggered to halt the market on January 4 and 7, and shook the investors' confidences. During that week, the SSE Composite Index and the SZSE Component Index slumped for 9.97 % and 14.02 %, respectively (see Fig. 1). The tumble lasted until the end of January 2016. In January 27, the SSE Composite Index and the SZSE Component Index hit new rock bottom at 2638.30 points and 8986.52 points, respectively. Second, the Chinese Spring Festival is during the week of February 8 to 12, 2016, so the market was closed. The fluctuation occurred around the 'biggest' holiday of China again. On February 25, more than one thousand shares went to the down-limit (-10%); the SSE Composite Index and the SZSE Component Index fell 6.41 % to 2741.25 points and 7.34 % to 9551.08 points, respectively. On February 29, the SSE Composite Index and the SZSE Component Index tumbled to 2638.96 points and 9000.89 points, respectively, which are very close to the values on January 27. These two close valleys imply a revival and the big change in Fig. 6 also provides a sign. From Fig. 1, we see that the SSE Composite Index gained 9.53 % from March 11 (2810.31 points) to April 15 (3078.12 points), and the SZSE Component Index gained 14.63 % from March 11 (9363.41 points) to April 15 (10733.64 points).

5 Conclusion and Future Work

In summary, we computed some motif statistics, dual motif statistics, and the cut-distances for the correlation stock networks over the Chinese A-share Stocks. These measurements are motivated from the theory of graph limits. When a sequence of graphs $\{G_i\}$ converges, then all these (sequences of) quantities converge. In contraposition, the fluctuation of these quantities can be used to measure the fluctuation of the graphs. We observed significant changes on all three measures when the Chinese A-share Stocks changed dramatically.

In the future, we hope to establish a graphon-inspired real-time system to predict the future dramatic changes in the stock market. For example, we may set Δt to be one minute instead of one day and set the time intervals by 10 or 20 minutes. Of course, such tasks put a heavy burden on the computing, so fast sampling results like Theorem 1 are useful to estimate these measurements.

Acknowledgments. We would like to thank the referees for their helpful comments.

Appendix

There are 122 weeks between December 30, 2013 and April 29, 2016. The irregular data during 12 weeks are excluded due to Chinese holidays. Therefore, only 110 weekly correlation graphs were constructed. Here are the list of those 110 weeks.

Table 1. List of the weeks

1	2	3	4	5	6	7	8
12/30/2013 – 01/03/2014	01/06/2014 – 01/10/2014	01/13/2014 – 01/17/2014	01/20/2014 – 01/24/2014	01/27/2014 – 01/31/2014	02/10/2014 – 02/14/2014	02/17/2014 – 02/21/2014	02/24/2014 – 02/28/2014
9	10	11	12	13	14	15	16
03/03/2014 – 03/07/2014	03/10/2014 – 03/14/2014	03/17/2014 – 03/21/2014	03/24/2014 – 03/28/2014	03/31/2014 – 04/04/2014	04/07/2014 – 04/11/2014	04/14/2014 – 04/18/2014	04/21/2014 – 04/25/2014
17	18	19	20	21	22	23	24
05/05/2014 – 05/09/2014	05/12/2014 – 05/16/2014	05/19/2014 – 05/23/2014	05/26/2014 – 05/30/2014	06/02/2014 – 06/06/2014	06/09/2014 – 06/13/2014	06/16/2014 – 06/20/2014	06/23/2014 – 06/27/2014
25	26	27	28	29	30	31	32
06/30/2014 – 07/04/2014	07/07/2014 – 07/11/2014	07/14/2014 – 07/18/2014	07/21/2014 – 07/25/2014	07/28/2014 – 08/01/2014	08/04/2014 – 08/08/2014	08/11/2014 – 08/15/2014	08/18/2014 – 08/22/2014
33	34	35	36	37	38	39	40
08/25/2014 – 08/29/2014	09/01/2014 – 09/05/2014	09/08/2014 – 09/12/2014	09/15/2014 – 09/19/2014	09/22/2014 – 09/26/2014	10/13/2014 – 10/17/2014	10/20/2014 – 10/24/2014	10/27/2014 – 10/31/2014
41	42	43	44	45	46	47	48
11/03/2014 – 11/07/2014	11/10/2014 – 11/14/2014	11/17/2014 – 11/21/2014	11/24/2014 – 11/28/2014	12/01/2014 – 12/05/2014	12/08/2014 – 12/12/2014	12/15/2014 – 12/19/2014	12/22/2014 – 12/26/2014
49	50	51	52	53	54	55	56
01/05/2015 – 01/09/2015	01/12/2015 – 01/16/2015	01/19/2015 – 01/23/2015	01/26/2015 – 01/30/2015	02/02/2015 – 02/06/2015	02/09/2015 – 02/13/2015	03/02/2015 – 03/06/2015	03/09/2015 – 03/13/2015
57	58	59	60	61	62	63	64
03/16/2015 – 03/20/2015	03/23/2015 – 03/27/2015	03/30/2015 – 04/03/2015	04/06/2015 – 04/10/2015	04/13/2015 – 04/17/2015	04/20/2015 – 04/24/2015	04/27/2015 – 05/01/2015	05/04/2015 – 05/08/2015
65	66	67	68	69	70	71	72
05/11/2015 – 05/15/2015	05/18/2015 – 05/22/2015	05/25/2015 – 05/29/2015	06/01/2015 – 06/05/2015	06/08/2015 – 06/12/2015	06/15/2015 – 06/19/2015	06/22/2015 – 06/26/2015	06/29/2015 – 07/03/2015
73	74	75	76	77	78	79	80
07/06/2015 – 07/10/2015	07/13/2015 – 07/17/2015	07/20/2015 – 07/24/2015	07/27/2015 – 07/31/2015	08/03/2015 – 08/07/2015	08/10/2015 – 08/14/2015	08/17/2015 – 08/21/2015	08/24/2015 – 08/28/2015
81	82	83	84	85	86	87	88
09/07/2015 – 09/11/2015	09/14/2015 – 09/18/2015	09/21/2015 – 09/25/2015	10/12/2015 – 10/16/2015	10/19/2015 – 10/23/2015	10/26/2015 – 10/30/2015	11/02/2015 – 11/06/2015	11/09/2015 – 11/13/2015
89	90	91	92	93	94	95	96
11/16/2015 – 11/20/2015	11/23/2015 – 11/27/2015	11/30/2015 – 12/04/2015	12/07/2015 – 12/11/2015	12/14/2015 – 12/18/2015	12/21/2015 – 12/25/2015	12/28/2015 – 01/01/2016	01/11/2016 – 01/15/2016
97	98	99	100	101	102	103	104
01/18/2016 – 01/22/2016	01/25/2016 – 01/29/2016	02/01/2016 – 02/05/2016	02/15/2016 – 02/19/2016	02/22/2016 – 02/26/2016	02/29/2016 – 03/04/2016	03/07/2016 – 03/11/2016	03/14/2016 – 03/18/2016
105	106	107	108	109	110		
03/21/2016 – 03/25/2016	03/28/2016 – 04/01/2016	04/04/2016 – 04/08/2016	04/11/2016 – 04/15/2016	04/18/2016 – 04/22/2016	04/25/2016 – 04/29/2016		

References

1. Alon, U.: Network motifs: theory and experimental approaches. Nat. Rev. Genet. **8**, 450–461 (2007)
2. Alon, N., Naor, A.: Approximating the cut-norm via Grothendieck's inequality. SIAM J. Comput. **35**(4), 787–803 (2006)
3. Bhuiyan, M., Rahman, M., Rahman, M., Hasan, M.: GUISE: uniform sampling of graphlets for large graph analysis. In: Proceedings of IEEE ICDM 2012, pp. 91–100. IEEE Press (2012)
4. Birmelé, E.: Detecting local network motifs. Electron. J. Statist. **6**, 908–933 (2012)
5. Boginski, V., Butenko, S., Pardalos, P.M.: Statistical analysis of financial networks. Comput. Statist. Data Anal. **48**, 431–443 (2005)
6. Borgs, C., Chayes, J., Lovász, L., Sós, V.T., Vesztergombi, K.: Counting graph homomorphisms. In: Klazar, M., Kratochvil, J., Loebl, M., Matoušek, J., Valtr, P., Thomas, R. (eds.) Topics in Discrete Mathematics. Algorithms and Combinatorics, vol. 26, pp. 315–371. Springer, Heidelberg (2006)
7. Borgs, C., Chayes, J.T., Lovász, L., Sós, V.T., Vesztergombi, K.: Convergent sequences of dense graphs I: subgraph frequencies, metric properties and testing. Adv. Math. **219**, 1801–1851 (2008)
8. Borgs, C., Chayes, J.T., Lovász, L., Sós, V.T., Vesztergombi, K.: Convergent sequences of dense graphs II. Multiway cuts and statistical physics. Ann. Math. **176**, 151–219 (2012)
9. Chernoff, H.: A note on an inequality involving the normal distribution. Ann. Probab. **9**, 533–535 (1981)
10. Frieze, A., Kannan, R.: Quick approximation to matrices and applications. Combinatorica **19**, 175–220 (1999)
11. Huang, W.-Q., Zhuang, X.-T., Yao, S.: A network analysis of the Chinese stock market. Physica A **388**, 2956–2964 (2009)
12. Juszczyszyn, K., Kazienko, P., Musiał, K.: Local topology of social network based on motif analysis. In: Lovrek, I., Howlett, R.J., Jain, L.C. (eds.) KES 2008. LNCS (LNAI), vol. 5178, pp. 97–105. Springer, Heidelberg (2008)
13. Lovász, L.: Large Networks and Graph Limits. American Mathematical Society, Providence (2012)
14. Mantegna, R.N.: Hierarchical structure in financial markets. Eur. Phys. J. B **11**, 193–197 (1999)
15. Milo, R., Shen-Orr, S., Itzkovitz, S., Kashtan, N., Chklovskii, D., Alon, U.: Network motifs: simple building blocks of complex network. Science **298**, 824–827 (2002)
16. Namakia, A., Shirazi, A.H., Raei, R., Jafari, G.R.: Network analysis of a financial market based on genuine correlation and threshold method. Physica A **390**, 3835–3841 (2011)
17. Onnela, J.-P., Kaski, K., Kertész, J.: Clustering and information in correlation based financial networks. Eur. Phys. J. B **38**, 353–362 (2004)
18. Onnela, J.-P., Saramäki, J., Kertész, J., Kaski, K.: Intensity and coherence of motifs in weighted complex networks. Phys. Rev. E **71**, 065103(R) (2005)
19. Sporns, O., Kötter, R.: Motifs in brain networks. PLoS Biol. **2**(11), e:369, 1910–1918 (2004)
20. Wang, P., Lui, J.C.S., Ribeiro, B., Towsley, D., Zhao, J., Guan, X.: Efficiently estimating motif statistics of large networks. ACM Trans. Knowl. Discov. Data (TKDD) **9**(2), article 8 (2014)

Max Celebrity Games

C. Àlvarez[✉] and A. Messeguè

ALBCOM Research Group, Computer Science Department, UPC, Barcelona, Spain
alvarez@cs.upc.edu, arniszt@gmail.com

Abstract. We introduce *Max celebrity games* a new variant of Celebrity games defined in [4]. In both models players have weights and there is a critical distance β as well as a link cost α. In the max celebrity model the cost of a player depends on the cost of establishing links to other players and on the maximum of the weights of those nodes that are farther away than β (instead of the sum of weights as in celebrity games). The main results for $\beta > 1$ are that: computing a best response for a player is NP-hard; the optimal social cost of a celebrity game depends on the relation between α and w_{max}; NE always exist and NE graphs are either connected or a set of $r \geq 1$ connected components where at most one of them is not an isolated node; for the class of connected NE graphs we obtain a general upper bound of $2\beta + 2$ for the diameter. We also analyze the price of anarchy (PoA) of connected NE graphs and we show that there exist games Γ such that $\mathrm{PoA}(\Gamma) = \Theta(n/\beta)$; modifying the cost of a player we guarantee that all NE graphs are connected, but the diameter might be $n - 1$. Finally, when $\beta = 1$, computing a best response for a player is polynomial time solvable and the PoA $= O(w_{max}/w_{min})$.

1 Introduction

The increasing use of Internet and social networks, has motivated a great interest to model theoretically their behavior. Fabrikant et al. [15] proposed a game-theoretic model of network creation (NCG) as a simple tool to analyze the creation of Internet as a decentralized and non-cooperative communication network among players (the network nodes).

In this model the goal of each player is to have, in the resulting network, all the other nodes as close as possible paying a minimum cost. It is assumed that: all the players have the same interest (all-to-all communication pattern with identical weights); the cost of being disconnected is infinite; and the links to other nodes paid by one node can be used by others. Formally, a game Γ in this seminal model is defined as a tuple $\Gamma = \langle V, \alpha \rangle$, where V is the set of n nodes and α the cost of establishing a link. A strategy for player $u \in V$ is a subset $S_u \subseteq V - \{u\}$, the set of players for which player u pays for establishing a link. The n players and their joint strategy choices $S = (S_u)_{u \in V}$ create an undirected graph $G[S]$. The cost function for each node u under strategy S is defined by $c_u(S) = \alpha|S_u| + \sum_{v \in V} d_{G[S]}(u, v)$

© Springer International Publishing AG 2016
A. Bonato et al. (Eds.): WAW 2016, LNCS 10088, pp. 88–99, 2016.
DOI: 10.1007/978-3-319-49787-7_8

where $d_{G[S]}(u, v)$ is the distance between nodes u and v in graph $G[S]$. By changing the cost function to $c_u(S) = \alpha|s_u| + \max\{d_{G[S]}(u, v)|v \in V\}$ as proposed in [13] one obtains the max game model.

From here on several versions have been considered to make the model a little more realistic. For different variants we refer the interested reader to [1–3, 6, 9–14, 16, 17, 19] among others.

In Internet as well as in social networks not all the nodes have the same importance. It seems natural to consider nodes with different relevance weights. In such a setting, the cost of being far (even if connected) from high-weight nodes should be greater than the cost of being far from low-weight nodes. In [4] we introduce *celebrity games* with the aim to study the combined effect of having players with different weights that share a common distance bound.

In celebrity games the cost of a player has two components. The first one is the cost of the links established by the node. The second one is the sum of the weights of those nodes that are farther away than the critical distance. Formally, a celebrity game is defined by $\Gamma = \langle V, (w_u)_{u \in V}, \alpha, \beta \rangle$, where V is a set of nodes with weights $(w_u)_{u \in V}$, α is the cost of establishing a link and β establishes the desirable distance bound. The cost function for each node is defined by $c_u(S) = \alpha|S_u| + \sum_{\{v|d_{G[S]}(u,v)>\beta\}} w_v$.

In this paper we extend the study initiated in [4]: we define a max version of the celebrity games that we name *max celebrity games* and we analyze the structure and quality of their Nash equilibria. From now on, let us refer to celebrity games as *sum celebrity games*. In the max celebrity model the cost of a player takes into account the maximum of the weights (worst-case) of those nodes that are farther away than the critical distance, instead of the sum of weights (average-case). The cost function is formally defined by $c_u(S) = \alpha|S_u| + \max_{\{v|d_{G[S]}(u,v)>\beta\}} w_v$. Intuitively, the goal of each player in max celebrity games is to buy as few links as possible in order to have the high-weighted nodes closer to the given critical distance. Observe that if the cost of establishing links is higher than the benefit of having close a node (or set of nodes), players might rather prefer to stay either far or even disconnected from it.

Observe that the main feature of both, sum and max celebrity games, is the combination of bounded distance with players having different weights. Even though heterogeneous players have been considered in NCG under bilateral contracting [5, 18], and the notion of bounded distance has been studied in [8], to the best of our knowledge sum celebrity games is the first model that studies how a common critical distance, different weights, and a link cost, altogether affect the individual preferences of the players. Furthermore, max celebrity games is the first model that focuses on how the maximum weight of those nodes that are farther than β affects the creation of graphs.

In this paper we analyze the structure of Nash equilibrium (NE) graphs of max celebrity games and their quality with respect to the optimal strategies. To do so we address the cases $\beta > 1$ and $\beta = 1$, separately. For $\beta > 1$, every player u has to choose for each non-edge (u, v) between paying the maximum of the weights of the nodes with distance to u greater than β, or buying the link (u, v) and paying α for the link minus the maximum of the weights of those nodes whose distance to u will become less or equal than the critical distance β. While for $\beta = 1$, each player u has to decide for every non-edge (u, v) of the graph to pay either α for the link or at least w_v (the weight of the non-adjacent node v).

For the general case $\beta > 1$ our results can be summarized as follows: computing a best response for a player is NP-hard; the optimal social cost of a celebrity game Γ depends on the relation between α and the maximum weight w_{max}; NE always exist and NE graphs are either connected or a set of $r \geq 1$ connected components where at most one of them is not an isolated node; for the class of connected NE graphs we obtain a general upper bound of $2\beta + 2$ for the diameter; we also analyze the quality of connected NE graphs and we show that there exist max celebrity games such that $\text{PoA}(\Gamma) = \Theta(n/\beta)$; we consider a variation of the cost of the player in order to avoid non-connected NE graphs.

Finally, for the particular case $\beta = 1$, we show that computing a best response for a player is polynomial time solvable and that the $\text{PoA} = O(w_{max}/w_{min})$.

The paper is organized as follows. In Sect. 2 we introduce the basic definitions and the max celebrity model. In Sect. 3 we study the fundamental properties of optimal graphs and NE graphs. Section 4 studies for $\beta > 1$ the quality of connected NE graphs and considers a modification of the cost of a player in order to guarantee connected NE graphs. In Sect. 5 we study the complexity of the best response problem and the PoA for the case $\beta = 1$. Finally, in Sect. 6 we give an outline of the main differences between max and sum celebrity models.

2   The Model

We use standard notation for graphs and strategic games. All the graphs in the paper are undirected unless explicitly said otherwise. For a graph $G = (V, E)$ and $u, v \in E$, $d_G(u, v)$ denotes the distance, i.e. the length of a shortest path, from u to v in G. The *diameter* of a vertex $u \in V$, $diam_G(u)$, is defined as $diam_G(u) = \max_{v \in V}\{d_G(u, v)\}$ and the *diameter* of G, $diam(G)$, is defined as usual as $diam(G) = \max_{v \in V}\{diam_G(v)\}$. An *orientation* of an undirected graph is an assignment of a direction to each edge, turning the initial graph into a directed graph.

For a weighted set $(V, (w_u)_{u \in V})$ we extend the weight function to subsets in the following way. For $U \subseteq V$, $w(U) = \max_{u \in U}\{w_u\}$. Furthermore, we set $w_{max} = \max_{u \in V}\{w_u\}$ and $w_{min} = \min_{u \in V}\{w_u\}$.

Definition 1. *A max celebrity game Γ is defined by a tuple $\langle V, (w_u)_{u \in V}, \alpha, \beta \rangle$ where: $V = \{1, \ldots, n\}$ is the set of players, for each player $u \in V$; $w_u > 0$ is the weight of player u; $\alpha > 0$ is the cost of establishing a link and β, $1 \leq \beta \leq n - 1$, is the critical distance.*

A strategy for player u is a subset $S_u \subseteq V - \{u\}$ denoting the set of players for which player u pays for establishing a direct link. A strategy profile for Γ is a tuple $S = (S_1, S_2, \ldots, S_n)$ defining a strategy for each player. For a strategy profile S, the associated outcome graph is the undirected graph $G[S]$ which is defined by $G[S] = (V, \{\{u, v\} | u \in S_v \vee v \in S_u\})$.

For a strategy profile $S = (S_1, S_2, \ldots, S_n)$, *the cost function of player* u, *denoted by* c_u, *is defined as* $c_u(S) = \alpha|S_u| + W_u$ *where* $W_u = \max_{\{v|d_{G[S]}(u,v)>\beta\}}\{w_v\}$. *And as usual, the* social cost *of a strategy profile* S *in* Γ *is defined as* $C(S) = \sum_{u \in V} c_u(S)$. *The* social cost *of a graph* G *in* Γ *is defined analogously as* $C(G) = \alpha|E(G)| + \sum_{u \in V(G)} W_u$.

Observe that, even though a link might be established by only one player, we consider the outcome graph as an undirected graph, assuming that once a link is bought the link can be used in both directions. In our definition we have considered a general case in which players may have different weights and defined the cost function through properties of the undirected graph created by the strategy profile. The player's cost function takes into account two components: the cost of establishing links and the maximum of the weights of the players that are at a distance greater than the critical distance β.

In the remaining of the paper, we assume that, for $\Gamma = \langle V, (w_u)_{u \in V}, \alpha, \beta \rangle$, the parameters verify the required conditions. Furthermore, unless specifically stated, we consider $\beta > 1$, the case $\beta = 1$ will be studied in Sect. 5. We use the following notation, for a game $\Gamma = \langle V, (w_u)_{u \in V}, \alpha, \beta \rangle$, $n = |V|$. We denote by $\mathcal{S}(u)$ the set of strategies for player u and by $\mathcal{S}(\Gamma)$ the set of strategy profiles of Γ.

As usual, for a strategy profile S and a strategy S'_u for player u, (S_{-u}, S'_u) represents the strategy profile in which S_u is replaced by S'_u while the strategies of the other players remain unchanged. The *cost difference* $\Delta(S_{-u}, S'_u)$ is defined as $\Delta(S_{-u}, S'_u) = c_u(S_{-u}, S'_u) - c_u(S)$. Observe that, if $\Delta(S_{-u}, S'_u) < 0$, player u has an incentive to deviate from S_u. A best response to $S \in \mathcal{S}(\Gamma)$ for player u is a strategy $S'_u \in \mathcal{S}(u)$ minimizing $\Delta(S_{-u}, S'_u)$. Let us remind the definition of Nash equilibrium.

Definition 2. *Let* $\Gamma = \langle V, (w_u)_{u \in V}, \alpha, \beta \rangle$ *be a max celebrity game. A strategy profile* $S = (S_1, S_2, \ldots, S_n)$ *is a* Nash equilibria *of* Γ *if no player has an incentive to deviate from his strategy. Formally, for a player* u *and each strategy* $S'_u \in \mathcal{S}(u)$, $\Delta(S_{-u}, S'_u) \geq 0$.

We denote by $\mathrm{NE}(\Gamma)$ the set of Nash equilibria of a game Γ. We use the term NE to refer to a strategy profile $S \in \mathrm{NE}(\Gamma)$. We say that a graph G is a NE *graph* if there is $S \in \mathrm{NE}(\Gamma)$ so that $G = G[S]$.

We denote by $\mathsf{opt}(\Gamma)$ the minimum value of the social cost, i.e. $\mathsf{opt}(\Gamma) = \min_{S \in \mathcal{S}(\Gamma)} C(S)$. We denote by $\mathrm{OPT}(\Gamma)$ the set of optimum strategy profiles of Γ w.r.t. the social cost, that is, for $S \in \mathrm{OPT}(\Gamma)$, $C(S) = \mathsf{opt}(\Gamma)$. We use the term OPT *strategy profile* to refer to a $S \in \mathrm{OPT}(\Gamma)$.

It is worth observing that: for $S \in \mathrm{NE}(\Gamma)$, it never happens that $v \in S_u$ and $u \in S_v$, for any $u, v \in V$; a NE graph G can be the outcome of several strategy profiles and not all the orientations of a NE graph G are NE.

In the following we make use of some particular outcome graphs on n vertices: I_n, the independent set; and ST_n a star graph, i.e. a tree in which one of the vertices, the *central* vertex, is connected to all the other $n - 1$ vertices.

We define the Price of Anarchy and the Price of Stability as usual.

Definition 3. *Let Γ be a max celebrity game. The* Price of Anarchy *of Γ is defined as $PoA(\Gamma) = \max_{S \in NE(\Gamma)} C(S)/\mathsf{opt}(\Gamma)$ and the* Price of Stability *of Γ is defined as $PoS(\Gamma) = \min_{S \in NE(\Gamma)} C(S)/\mathsf{opt}(\Gamma)$*

The explicit reference to Γ will be dropped whenever Γ is clear from the context. We will refer to $NE(\Gamma)$, $\mathsf{opt}(\Gamma)$, $PoA(\Gamma)$, and $PoS(\Gamma)$ by NE, opt, PoA and PoS respectively.

Our first result shows that the computation of a best response in max celebrity games is a NP-hard problem for $\beta \geq 2$. The proof consists in a reduction from the Dominating Set problem. The problem becomes polynomial time computable for $\beta = 1$ as we show in Sect. 5.

Proposition 1. *Computing a best response for a player to a strategy profile in a max celebrity game is* NP-hard *even when $\beta = 2$.*

3   Social Optimum and Nash Equilibrium

In this section we analyze some properties of the opt and NE strategy profiles in max celebrity games. We start by giving bounds for opt that depend on the existence of one or more than one connected components.

Proposition 2. *Let $\Gamma = \langle V, (w_u)_{u \in V}, \alpha, \beta \rangle$ be a max celebrity game. We have that $2\alpha(n-1) \geq \mathsf{opt}(\Gamma) \geq \min\{\alpha(n-1), w_{max}(n-1) + w_{min}\}$.*

Proof. Let $S \in \mathrm{OPT}(\Gamma)$ and let $G = G[S]$. Let $G_1, ..., G_r$ be the connected components of G and let $V_i = V(G_i), k_i = |V_i|$, and $W_i = w(V_i)$, for $1 \leq i \leq r$. Assume w.l.o.g that $W_1 \geq W_2 \geq ... \geq W_r$. Observe that the social cost of a disconnected graph can be expressed as the sum of the social cost of the connected components plus the additional contribution of the pairs of vertices that lie in different components. Each connected component must be a tree of diameter at most β, otherwise a strategy profile with smaller social cost could be obtained by replacing the connections on V_i by such a tree. W.l.o.g we can assume that, for $1 \leq i \leq r$, the $i-$th connected component is a star graph ST_{k_i} on k_i vertices. Recall that $C(ST_{k_i}) = \alpha(k_i - 1)$, thus $C(G) = \sum_{i=1}^{r} \alpha(k_i - 1) + \sum_{i=1}^{r} k_i (\max_{j \neq i}\{W_j\}) = \alpha(n - r) + nW_1 - k_1(W_1 - W_2)$.

Notice that if for some $i > 1$, the i-th connected component is not an isolated node, then the node with maximum weight in this connected component can be moved to G_1. By preserving the connectivity and structure (a star) of G_1, the social cost of the resulting graph is strictly smaller than the cost of the original G. This implies that for every $i > 1$, $k_i = 1$. Hence, $C(G) = \alpha(n-r) + (r-1)W_1 + (n-r+1)W_2$.

If $r = 1$ then $C(G) = \alpha(n-1)$ and we are done. Otherwise, if $r > 1$, then we have the inequality $C(ST_n) \geq C(G)$. This implies that $\alpha \geq \frac{1}{r-1}(W_1(r-1) + W_2(n-r+1)) \geq W_1$.

Then we get the following results. First: $C(G) \geq W_1(n-r) + (r-1)W_1 + (n-r+1)W_2 \geq (n-1)W_1 + W_2 \geq (n-1)w_{max} + w_{min}$.

Secondly, using that $r > 1$: $C(G) = \alpha(n-r) + (r-1)W_1 + (n-r+1)W_2 \leq \alpha(n-r) + (r-1)\alpha + (n-r+1)\alpha \leq 2\alpha(n-1)$.

The relationship between α and w_{max} determines partially the topology of the NE graphs. As one can expect, if $\alpha > w_{max}$, no player has incentive to establish a link then the independent set is the unique NE graph. Otherwise, any NE graph can be connected or disconnected.

Proposition 3. *Every max celebrity game* $\Gamma = \langle V, (w_u)_{u \in V}, \alpha, \beta \rangle$ *has a* NE. *Furthermore, when* $\alpha \leq w_{max}$, ST_n *is a* NE *graph, and when* $\alpha \geq w_{max}$, I_n *is a* NE *graph. If* $\alpha > w_{max}$, *then* I_n *is the unique* NE *graph.*

Proof. When $\alpha \leq w_{max}$ let us show that ST_n is a NE graph. Let u_{max} a node with maximum weight and we suppose that it is the central node of the star. If $S_{u_{max}} = \emptyset$ and for every node $v \neq u_{max}$, $S_v = \{u_{max}\}$, then u_{max} cannot improve its actual cost since it is exactly 0. Moreover, the other nodes can only delete the edge to u_{max}. Since such deviation has a cost increment of $-\alpha + w_{max} \geq 0$, then we are done.

When $\alpha \geq w_{max}$, let us show that I_n is a NE graph. Let S be the empty strategy profile, $I_n = G[S]$. Notice that for any player u, if $S'_u \neq \emptyset$, then $\Delta(S_{-u}, S'_u) \geq |S'_u|\alpha - w_{max} \geq (|S'_u| - 1)w_{max} \geq 0$. Finally, if $\alpha > w_{max}$ it is easy to see that the unique NE graph is I_n. Let us suppose that there exist $u, v \in V$ such that $v \in S_u$. If $S'_u = S_u - \{v\}$, then $\Delta(S_{-u}, S'_u) \leq -\alpha + w_{max} < 0$. Hence, if $G \neq I_n$, then G is not a NE graph.

Corollary 1. *Let* $\Gamma = \langle V, (w_u)_{u \in V}, \alpha, \beta \rangle$ *be a max celebrity game. Then,* $PoS(\Gamma) = 1$ *for* $\alpha \leq w_{max}$ *and* $PoS(\Gamma) < 2$ *for* $\alpha > w_{max}$.

In particular, even in the case that $\alpha < w_{max}$, it can be shown that there exist max celebrity games where I_n is a NE graph. Indeed consider a game with $n \geq 2$ and weights defined as $w_i = (i - 1)\alpha$ for $i > 1$ and $w_1 = \alpha$. Then, clearly $\alpha < w_{max}$ and I_n is a NE graph.

Furthermore, for every integer $1 < r \leq n$, there exists non-connected max celebrity games with exactly r different connected components. Moreover, the only connected component that can have more than one node is the one that contains a node with weight w_{max}.

Proposition 4. *Any* NE *graph distinct from* I_n *has at most one non-trivial connected component. Moreover, for every integer* $r \geq 2$ *there exists a max celebrity game having a* NE *graph with exactly* r *connected components.*

Proof. (Sketch) For the first part, let $G_1, ..., G_r$ be the connected components of a NE graph. Assume that a node with the maximum weight is in G_1. If for some $i > 1$, $|G_i| > 1$, then there exist $u, v \in V(G_i)$ such that $u \in S_v$. In this case, v can strictly decrease its cost deleting this edge because the node with maximum weight is still at distance greater than β, contradicting the fact that G is a NE graph.

For the second part we distinguish two cases: $r \geq 3$ and $r = 2$. For the first case, let $n = r + 1$, $V = \{v_0, v_1, ..., v_r\}$, $E = \{\{v_0, v_1\}\}$ with $S_{v_1} = \{v_0\}$, $S_{v_i} = \emptyset$ for $i \neq 1$, as depicted in the figure below. For the weights consider $w_{v_0} = w_1$ and

$w_{v_i} = w_2$ for all $i \geq 1$, with $w_1 > w_2$, $w_1 - w_2 = \alpha$ and $\alpha \geq w_1/(n-1)$, $w_2/(n-2)$. We have that this configuration is a NE.

For the case $r = 2$ see the figure below. It is not hard to see that this configuration is also a NE.

4   The Price of Anarchy

Observe that there exist max celebrity games Γ with $\alpha \leq w_{max}$ having disconnected NE graphs with high social cost in comparison with the optimum. Indeed, consider the example given in Proposition 4 with $w_2 = \frac{w_1(n-2)}{(n-1)}$. The cost of this NE graph is $w_1(n-1) + \frac{w_1(n-2)}{n-1}$ and combining it with opt $\leq 2\alpha(n-1)$, we get the bound $PoA(\Gamma) \geq (n-1)/2$. Hence, we focus on the study of the PoA for connected NE graphs. Since the restriction $\alpha \leq w_{max}$ by itself does not exclude the possibility of having non-connected NE graphs, we study the PoA of connected equilibria from two different perspectives: first, we analyze the worst case among all connected NE graphs; second, we introduce a slight modification of the player's cost function in order to guarantee connectivity in the class of NE graphs. Whenever we consider the class of connected NE graphs we compare the social cost of such equilibria with the optimum value among the connected graphs, $opt(\Gamma) = \alpha(n-1)$.

4.1   PoA and Diameter of Connected NE Graphs

In this subsection we analyze the quality and structure of equilibria in terms of the parameters that define the max celebrity games. Our next result indicates that the price to pay for the anarchy is low when α is close to w_{max}.

Proposition 5. *For every max celebrity game* $\Gamma = \langle V, (w_u)_{u \in V}, \alpha, \beta, \rangle$, $PoA(\Gamma) \leq 2(w_{max}/\alpha)$.

Proof. Let S be a NE of Γ and let $G = G[S] = (V, E)$. Then, no player has an incentive to deviate from S. In particular, for each $u \in V$ we have that $0 \leq \Delta(S_{-u}, \emptyset) = -\alpha|S_u| + W'_u - W_u$ where $W_u = \max_{\{x \mid d_G(u,x) > \beta\}} w_x$ and $W'_u = \max_{\{x \mid d_{G[(S_{-u}, \emptyset)]}(u,x) > \beta\}} w_x$. By adding for each $u \in V$ the corresponding inequalities, we have that $0 \leq \sum_{u \in V}(-\alpha|S_u| + W'_u - W_u) = -\alpha|E| + \sum_{u \in V} W'_u - \sum_{u \in V} W_u$.

Therefore, $C(G) = \alpha|E| + \sum_{u \in V} W_u \leq \sum_{u \in V} W'_u \leq nw_{max}$ and we can conclude that $PoA(\Gamma) \leq \frac{nw_{max}}{\alpha(n-1)} \leq 2\frac{w_{max}}{\alpha}$.

The diameter of NE graphs depends directly on the critical distance β.

Proposition 6. *Let $\Gamma = \langle V, (w_u)_{u \in V}, \alpha, \beta \rangle$ be a max celebrity game. In a* NE *graph G for Γ, $diam(G) \le 2\beta + 2$.*

Proof. Let $S \in \text{NE}(\Gamma)$ such that $G = G[S]$ is connected. Assume that the node u satisfies that $d_G(u, u_{max}) > \beta$ and $|S_u| > 0$. Then u has incentive to break any of its bought links because after doing so, u_{max} will still remain inside the complementary of the ball of radius β centered at u. Next, assume that $diam(u_{max}) \ge \beta + 2$. Let $u_{max}, u_1, u_2, ..., u_{\beta+2}$ be a path. Then, either $u_{\beta+1} \in S_{u_{\beta+2}}$ or $u_{\beta+2} \in S_{u_{\beta+1}}$. Therefore, since both $u_{\beta+1}, u_{\beta+2}$ are at distances greater than β from u_{max}, G cannot be a NE. This proves that $diam(u_{max}) \le \beta + 1$ in any connected NE and, as a consequence, that $diam(G) \le 2\beta + 2$.

Let us provide for a NE graph G, a bound on the contribution of the *weight component* of the social cost of G, $W(G, \beta) = \sum_{\{u \in V(G)\}} W_u$. The following lemma is a reformulation of a similar result that can be found in [4] using a cleaner and simpler argument.

Lemma 1. *Let $\Gamma = \langle V, (w_u)_{u \in V}, \alpha, \beta \rangle$ be a max celebrity game. In a* NE *graph G for Γ, $W(G, \beta) = O(\alpha n^2 / \beta)$.*

Proof. Let S be a NE and $G = G[S]$ be a connected NE graph. Let $u \in V$ be any node in V. Consider the sets $A_i(u) = \{v \mid d_G(u, v) = i\}$. Define for $i = 1, ..., k$, $C_i = \{v \in V \mid (i-1)(\beta - 1) \le d_G(u, v) < i(\beta - 1)\} = \cup_{(i-1)(\beta-1) \le j < i(\beta-1)} A_j(u)$ with k such that $\cup_{i=1}^{k} C_i = V(G)$. By the pigeonhole principle, for each $i = 1, ..., k$ there exists at least one subindex, call it $j(i)$, for which $(i-1)(\beta - 1) \le j(i) < i(\beta - 1)$ and $|A_{j(i)}(u)| \le |C_i|/(\beta - 1)$. In this way, for any node $v \in V(G)$, let $S'_v = (S_v \cup_{i=1}^{k} A_{j(i)}(u)) - \{v\}$ and let $G' = G[S_{-v}, S'_v]$. By construction, $diam_{G'}(v) \le \beta$. Therefore, as S is a NE, we have $0 \le \Delta(S_{-v}, S'_v) \le \alpha \sum_{i=1}^{k} \frac{|C_i|}{\beta-1} - W_v = \alpha \left(\frac{n-1}{\beta-1} \right) - W_v$.

Thus, $W(G, \beta) \le \frac{n(n-1)\alpha}{\beta-1} = O\left(\frac{\alpha n^2}{\beta} \right)$.

Using the same technique to provide a bound for the number of edges in NE graphs for a sum celebrity games (Proof of Lemma 4 of [4]), we obtain the following result.

Lemma 2. *Let $\Gamma = \langle V, (w_u)_{u \in V}, \alpha, \beta \rangle$ be a max celebrity game. In a* NE *graph G for Γ, $|E(G)| \le n - 1 + \frac{3n^2}{\beta}$.*

Corollary 2. *For every max celebrity game $\Gamma = \langle V, (w_u)_{u \in V}, \alpha, \beta, \rangle$, $PoA(\Gamma) = O(n/\beta)$*

Proposition 7. *For every $n > \beta > 1$, there exists a max celebrity game $\Gamma = \langle V, (w_u)_{u \in V}, \alpha, \beta, \rangle$ such that $PoA(\Gamma) = \Omega(n/\beta)$.*

Proof. Given n, let k and r be such that $n-1 = k\lfloor \beta \rfloor + r$, $k \ge 3$ and $0 \le r < \lfloor \beta \rfloor$. Let $V = \{u\} \cup \left(\cup_{i=1}^{k} \{u_{i,j} \mid 1 \le j \le \lfloor \beta \rfloor\} \right) \cup \{u_{k+1,1}, u_{k+1,2}, ..., u_{k+1,r}\}$. We then define $w_u = W$ and $w_{u_{i,j}} = w$ with W, w such that $w = (k-2)\alpha$ and $W > n\alpha$.

In this way we consider the configuration S defined by the relations $S_u = \emptyset$, $S_{u_{i,j}} = \{u_{i,j-1}\}$ for $j \geq 2$ and $S_{u_{i,1}} = \{u\}$ for $i = 1, ..., k+1$, as depicted in the figure below. To prove that $S \in \mathrm{NE}(\Gamma)$ we see that the cost difference associated to any deviation is not negative.

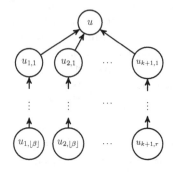

Clearly, u has no incentive of deviating his strategy because his cost is zero. Let us prove that any other node $u_{i,j}$ has no incentive in deviating from its current strategy. We say that a node v is covered with respect a node v' if v is at a distance at most β from v'. We have three cases:

1. *The deviation is such that all nodes are covered with respect $u_{i,j}$.* In this situation the cost difference is $l\alpha - w$. Notice that every node $u_{h,\lfloor\beta\rfloor}$ with $h \neq i$ can be reached only when a link from $u_{i,j}$ to the path formed by $u_{h,1}, u_{h,2}, ..., u_{h,\lfloor\beta\rfloor}$ is bought. Since initially $u_{i,j}$ has bought one link this leads to the inequality $l \geq k-2$. Therefore $l\alpha - w \geq (k-2)\alpha - (k-2)\alpha = 0$.
2. *The deviation is such that u is uncovered with respect $u_{i,j}$.* In this situation, since $W > n\alpha$, the cost difference is $l\alpha - w + W \geq 0$, for $-1 \leq l \leq n-1$.
3. *The deviation is such that u is covered with respect $u_{i,j}$ but there is at least one node node of weight w uncovered with respect $u_{i,j}$.* Then the cost difference is $l\alpha$ for some integer l. The only negative value that l can take is -1, but in such case the configuration leaves u uncovered with respect $u_{i,j}$, a contradiction. Therefore, $l\alpha \geq 0$.

Hence, $S \in \mathrm{NE}(\Gamma)$ and $C(S) > (n-1)w = (n-1)(k-2)\alpha$. Using the bound for the social optimum $\mathrm{opt}(\Gamma) \leq 2\alpha(n-1)$ we have that $\mathrm{PoA}(\Gamma) \geq (k-2)/2$.

Theorem 1. *For every $n > \beta > 1$, there exists a max celebrity game $\Gamma = \langle V, (w_u)_{u \in V}, \alpha, \beta, \rangle$ such that $PoA(\Gamma) = \Theta(n/\beta)$.*

4.2 The PoA When the Connectivity of the NE Graphs Is Guaranteed

Let us consider a new cost function that excludes non-connected NE graphs. We define a *connected max celebrity game* Γ^{con} as a max celebrity game $\Gamma^{con} = \langle V, (w_u)_{u \in V}, \alpha, \beta, \rangle$, but now, the cost for each player $u \in V$ in strategy profile S is denoted by $c_u^{con}(S)$ and it is defined as follows: $c_u^{con}(S) = c_u(S)$, if $diam_{G[S]}(u) \leq n-1$; otherwise, $c_u^{con}(S) = \infty$. As usual, the social cost of a strategy profile S in Γ^{con} is defined as $C^{con}(S) = \Sigma_{u \in V} c_u^{con}(S)$. Since for any connected graph G, $C^{con}(G) = C(G) \geq \alpha(n-1)$, then we have that $\mathrm{opt}(\Gamma^{con}) = \alpha(n-1)$. Notice that the same tuple $\langle V, (w_u)_{u \in V}, \alpha, \beta, \rangle$ can define a max celebrity game as well as a connected max celebrity game. In order to distinguish one from the other, we denote by $\Gamma = \Gamma(\langle V, (w_u)_{u \in V}, \alpha, \beta, \rangle)$ the corresponding max celebrity game and by $\Gamma^{con} = \Gamma^{con}(\langle V, (w_u)_{u \in V}, \alpha, \beta, \rangle)$, the corresponding connected max celebrity game.

Proposition 8. *Let $\langle V, (w_u)_{u \in V}, \alpha, \beta \rangle$ be a tuple defining $\Gamma = \Gamma(\langle V, (w_u)_{u \in V}, \alpha, \beta \rangle)$ and $\Gamma^{con} = \Gamma^{con}(\langle V, (w_u)_{u \in V}, \alpha, \beta \rangle)$. Then, $\mathrm{NE}(\Gamma) \subsetneq \mathrm{NE}(\Gamma^{con})$ when we consider $\mathrm{NE}(\Gamma)$ restricted to connected graphs.*

Proof. Let $S \in \mathrm{NE}(\Gamma)$ be such that $G = G[S]$ is connected. Let u be a player, let S_u' be a deviation, and let $G' = G[(S_{-u}, S_u')]$. Let $\Delta(S_{-u}, S_u')$ and $\Delta^{con}(S_{-u}, S_u')$ be the corresponding increments in the games Γ and Γ^{con}, respectively. We have that $\Delta^{con}(S_{-u}, S_u') = \Delta(S_{-u}, S_u')$, if G' is connected. Otherwise, $\Delta^{con}(S_{-u}, S_u') = \infty$, $\Delta(S_{-u}, S_u') < \infty$. Therefore, $\Delta^{con}(S_{-u}, S_u') \geq \Delta(S_{-u}, S_u')$ and then, $\mathrm{NE}(\Gamma) \subseteq \mathrm{NE}(\Gamma^{con})$.

To see that the inclusion might be strict, let us consider that $V = \{u, v\}$, $v \in S_u$, and $S_v = \emptyset$. If $w_v > \alpha$, S is not a NE for Γ. On the other hand, independently of the weights of u, v, S is a NE for Γ^{con}.

Proposition 9. *There are connected max celebrity games that have NE graphs with diameter equal to $n-1$.*

Proof. Let $n = 2k+1$ be a positive integer and let $V = \{v, v_1, v_{-1}, v_2, v_{-2}, \ldots, v_k, v_{-k}\}$. Let S be the strategy profile defined by $v_1, v_{-1} \in S_v$ and $v_{i+1} \in S_{v_i}, v_{-(i+1)} \in S_{v_{-i}}$ for $i \leq k-1$ (see the figure below). Setting the weights $w_x \leq \alpha$ for all $x \in V$ and for any $\beta < (n-1)/4$ it is easy to see that the corresponding graph is indeed a NE.

The bounds on the PoA obtained for the class of connected NE graphs for max celebrity games also hold for connected max celebrity games. The proofs also work for this case.

Theorem 2. *The PoA for the connected max celebrity games satisfies:*

1. *For every connected max celebrity game $\Gamma^{con} = \Gamma^{con}(\langle V, (w_u)_{u \in V}, \alpha, \beta \rangle)$, $PoA(\Gamma^{con}) = O(n/\beta)$*
2. *For every $n > \beta > 1$, there exists a connected max celebrity game $\Gamma^{con} = \Gamma^{con}(\langle V, (w_u)_{u \in V}, \alpha, \beta \rangle)$ such that $PoA(\Gamma^{con}) = \Theta(n/\beta)$.*

5   Max Celebrity Games for $\beta = 1$

When $\beta = 1$, each player u has to decide for every non-edge (u, v) of the graph to pay either α for the link, or at least w_v. It is not difficult to show that the best response of a player can be computed by sorting the weights of the non-adjacent nodes and then, selecting the number of links to be added to the most weighted non-adjacent nodes.

Proposition 10. *The problem of computing a best response of a player to a strategy profile in max celebrity games is polynomial time solvable when $\beta = 1$.*

In the next result we show that the price to pay for the anarchy is low when w_{min} is close to w_{max}.

Theorem 3. *Let $\Gamma = \langle V, (w_u)_{u \in V}, \alpha, 1 \rangle$ be a max celebrity game. Then, $PoA(\Gamma) = O(w_{max}/w_{min})$.*

Proof. Let $S \in \text{OPT}(\Gamma)$ and $G = G[S] = (V, A)$. Let $X = \{v \in V \mid deg(v) = n - 1\}$ where $deg(v)$ means the degree of v in the undirected graph G. We have that $C(G) \geq \frac{1}{2}\alpha(n - 1)|X| + (n - |X|)w_{min}$. Hence, $C(G) \geq nw_{min}$, if $w_{min} \leq \frac{(n-1)}{2}\alpha$ and $C(G) \geq \binom{n}{2}\alpha$, otherwise. To prove the result we distinguish three cases:

First we see that if $w_{min} \leq \alpha(n-1)/2$, then $PoA(\Gamma) \leq w_{max}/w_{min}$. Indeed, let S be a NE of Γ and let $G = G[S] = (V, E)$. Using the same reasoning as in Proposition 5 we have that $C(G) = \sum_{u \in V}(|S_u|\alpha + \max_{\{x|d(u,x)>1\}}\{w_x\}) \leq nw_{max}$. Therefore, if $w_{min} \leq \alpha(n-1)/2$, then $PoA(\Gamma) \leq w_{max}/w_{min}$, as we wanted to see.

Now, let us see that $PoA(\Gamma) = 1$ for $w_{min} > (n-1)\alpha$. This is because if $G \neq K_n$ then there exists some $v \in V$ with $diam_G(v) > 1$. Then considering the deviation for player v that consists in adding links to all the remaining nodes from the graph we get a cost increment of $k\alpha - w$ for some $k > 0$ and $w \geq w_{min}$. Since $k \leq (n-1)$ then $k\alpha - w \leq (n-1)\alpha - w_{min} < 0$, a contradiction for G being a NE. Thus $G = K_n$ and hence the result.

Finally, we see that for $\frac{n-1}{2}\alpha < w_{min} \leq (n-1)\alpha$ then $PoA(\Gamma) \leq 3$. Indeed, let S be a NE and $G = G[S] = (V, A)$. For a given $u \in V$ such that $diam_G(u) > 1$, let v be such that $w_v = W_u$. If $w_v > (n-1)\alpha$ then buying from u all the links to the remaining nodes from $V - \{x \mid d_G(u, x) \leq 1\}$ yields a cost increment of at most $(n-1)\alpha - w_v < 0$, a contradiction with G being a NE. Therefore $PoA(\Gamma) \leq ((\binom{n}{2})\alpha + n(n-1)\alpha)/\binom{n}{2}\alpha = 3$.

6   Max Celebrity Games Vs Sum Celebrity Games

The main differences between max and sum celebrity games are that: for $\beta > 1$, in max model there exist other disconnected NE graphs than I_n; in connected NE graphs, PoA $= O(n/\beta)$ in both models, but this is tight for some max games; for $\beta = 1$, PoA $= O(w_{max}/w_{min})$ in max, while in sum PoA ≤ 2. Finally, max celebrity games are equivalent to the MaxBD games (see [7,8]) when $\alpha < w_{min}/(n-1)$ as they are sum celebrity games when $\alpha < w_{min}$. (See the proof of Proposition 8 in [4] and replace $\alpha < w_{min}$ by $\alpha < w_{min}/(n-1)$).

Acknowledgements. This work was partially supported by funds from the AGAUR of the Government of Catalonia under project ref. SGR 2014:1034 (ALBCOM). C. Àlvarez was partially supported by the Spanish Ministry for Economy and Competitiveness (MINECO) and the European Union (FEDER funds), under grants ref. TIN2013-46181-C2-1-R (COMMAS).

References

1. Albers, S., Eilts, S., Even-Dar, E., Mansour, Y., Roditty, L.: On Nash equilibria for a network creation game. In: SODA 2006, pp. 89–98 (2006)
2. Alon, N., Demaine, E.D., Hajiaghayi, M., Kanellopoulos, P., Leighton, T.: Basic network creation games. SIAM J. Discrete Math. **27**(2), 656–668 (2013)
3. Alon, N., Demaine, E.D., Hajiaghayi, M., Kanellopoulos, P., Leighton, T.: Correction: basic network creation games. SIAM J. Discrete Math. **28**(3), 1638–1640 (2014)
4. Àlvarez, C., Blesa, M.J., Duch, A., Messegué, A., Serna, M.: Celebrity games. Theor. Comput. Sci. **648**, 56–71 (2016)
5. Àlvarez, C., Serna, M.J., Fernàndez, A.: Network formation for asymmetric players and bilateral contracting. Theor. Comput. Syst. **59**(3), 397–415 (2016)
6. Bilò, D., Guala, L., Leucci, S., Proietti, G.: The max-distance network creation game on general host graphs. Theor. Comput. Sci. **573**, 43–53 (2015a)
7. Bilò, D., Guala, L., Proietti, G.: Bounded-distance network creation games. In: Goldberg, P.W. (ed.) WINE 2012. LNCS, vol. 7695, pp. 72–85. Springer, Heidelberg (2012)
8. Bilò, D., Guala, L., Proietti, G.: Bounded-distance network creation games. ACM Trans. Econ. Comput. **3**(3), 16 (2015b)
9. Brandes, U., Hoefer, M., Nick, B.: Network creation games with disconnected equilibria. In: Papadimitriou, C., Zhang, S. (eds.) WINE 2008. LNCS, vol. 5385, pp. 394–401. Springer, Heidelberg (2008)
10. Corbo, J., Parkes, D.C.: The price of selfish behavior in bilateral network formation. PODC **2005**, 99–107 (2005)
11. Cord-Landwehr, A., Lenzner, P.: Network creation games: think global – act local. In: Italiano, G.F., Pighizzini, G., Sannella, D.T. (eds.) MFCS 2015. LNCS, vol. 9235, pp. 248–260. Springer, Heidelberg (2015). doi:10.1007/978-3-662-48054-0_21
12. Demaine, E.D., Hajiaghayi, M.T., Mahini, H., Zadimoghaddam, M.: The price of anarchy in cooperative network creation games. ACM SIGecom Exch. **8**(2), 2 (2009)
13. Demaine, E.D., Hajiaghayi, M.T., Mahini, H., Zadimoghaddam, M.: The price of anarchy in network creation games. ACM Trans. Algorithms **8**(2), 13 (2012)
14. Ehsani, S., Fadaee, S.S., Fazli, M., Mehrabian, A., Sadeghabad, S.S., Safari, M., Saghafian, M.: A bounded budget network creation game. ACM Trans. Algorithms **11**(4), 34 (2015)
15. Fabrikant, A., Luthra, A., Maneva, E.N., Papadimitriou, C.H., Shenker, S.: On a network creation game. PODC **2003**, 347–351 (2003)
16. Lenzner, P.: On dynamics in basic network creation games. In: Persiano, G. (ed.) SAGT 2011. LNCS, vol. 6982, pp. 254–265. Springer, Heidelberg (2011)
17. Leonardi, S., Sankowski, P.: Network formation games with local coalitions. PODC **2007**, 299–305 (2007)
18. Meirom, E.A., Mannor, S., Orda, A.: Network formation games with heterogeneous players and the internet structure. EC **2014**, 735–752 (2014)
19. Nikoletseas, S.E., Panagopoulou, P.N., Raptopoulos, C., Spirakis, P.G.: On the structure of equilibria in basic network formation. Theor. Comput. Sci. **590**(C), 96–105 (2015)

Mining and Modeling Character Networks

Anthony Bonato[1]([⊠]), David Ryan D'Angelo[1],
Ethan R. Elenberg[2], David F. Gleich[3], and Yangyang Hou[3]

[1] Ryerson University, Toronto, Canada
[2] University of Texas at Austin, Austin, USA
elenberg@utexas.edu
[3] Purdue University, West Lafayette, USA

Abstract. We investigate social networks of characters found in cultural works such as novels and films. These *character networks* exhibit many of the properties of complex networks such as skewed degree distribution and community structure, but may be of relatively small order with a high multiplicity of edges. Building on recent work of Beveridge and Shan [4], we consider graph extraction, visualization, and network statistics for three novels: *Twilight* by Stephanie Meyer, Steven King's *The Stand,* and J.K. Rowling's *Harry Potter and the Goblet of Fire.* Coupling with 800 character networks from films found in the http://moviegalaxies. com/ database, we compare the data sets to simulations from various stochastic complex networks models including random graphs with given expected degrees (also known as the Chung-Lu model), the configuration model, and the preferential attachment model. Using machine learning techniques based on motif (or small subgraph) counts, we determine that the Chung-Lu model best fits character networks and we conjecture why this may be the case.

1 Introduction

Complex networks lie at the intersection of several disciplines and have found broad application within the study of social networks. In social networks, nodes represents agents, and edges correspond to some kind of social interaction such as friendship or following. For more on complex networks and on-line social networks, the reader is directed to the book [5] and the survey [6].

In the present paper, we consider social networks arising in the context of cultural works such as novels or movies. In these *character networks*, nodes represent characters in a specified fictional or non-fictional work such as a novel, script, biography, or story, with edges between characters determined by their interaction within the work. We also consider character networks as weighted graphs, where the weights are positive integers specifying the co-appearance or co-occurrence of character names within a specified range of the text or scenes

Research supported by grants from NSERC and Ryerson University; Gleich and Hou's work were supported by NSF CAREER Award CCF-1149756, IIS-1546488, CCF-093937, and DARPA SIMPLEX.

A. Bonato et al. (Eds.): WAW 2016, LNCS 10088, pp. 100–114, 2016.
DOI: 10.1007/978-3-319-49787-7_9

(such as being within fifteen words of each other; see [4]). Not surprisingly, character networks are typically of smaller order than many other types of complex networks. Nevertheless, they still exhibit many of the interesting features of complex networks including clearly defined community structure, with communities centered on the various protagonists of the story, skewed degree distributions, focused on the most important characters, and dynamics. Character networks defined over larger fictional universes, such as the Marvel Universe, even grow to over 10,000 nodes [1,12].

There is an emerging approach using the tools of graph theory and big data to mine and model character networks. This new topic reflects the ease of access of cultural works in electronic formats, and the efficacy of big data-theoretic algorithms. Our approach in this work is study new networks with these tools to replicate some of the findings as well as study network models of these data.

First, we wish to study the complexity of these character networks through graph mining. Our approach here is more a microscopic view of an individual work's network. We focus on three well known novels: *Twilight* by Stephanie Meyer, Steven King's *The Stand*, and J.K. Rowling's *Harry Potter and the Goblet of Fire*. Various complex network statistics, such as diameter and clustering coefficient, are presented along with centrality metrics (such as PageRank and betweenness, paralleling the approach of [4]) that predict the major characters within each book. See Sect. 2 for the methodology used, and Sect. 3 has a summary of our results.

The second part of our approach is to compare and contrast the character networks with several well known stochastic network models. Hence, in this approach, we take a broader, macroscopic view of the structure of a larger sample of character networks. Using motifs (that is, small subgraph counts), eigenvalues, and machine learning techniques, we develop an approach for model selection for character networks. The models considered were the configuration model, preferential attachment model, the Chung-Lu model for random graphs with given expected degree sequences, and the binomial random graph (as a control). The parameters of the models were chosen as to equal the number of nodes and average degree of the character network data sets. Model selection was conducted for the three novels described above, and also for a set of 800 networks arising from movies in the http://moviegalaxies.com/ database [15]. Our results show consistent selection of the Chung-Lu model as the most realistic, with a clear separation between the models. We will discuss possible interpretations and implications of our results in the final section.

We consider undirected graphs throughout the paper. For background on graph theory, the reader is directed to [22]. Additional background on machine learning can be found in [13,20].

1.1   Previous Work

Quantitative methods have now emerged as a modern tool for literary analysis. Literary theories are now supported, debated, and refuted based on data [10]. In recent work, Reagan et al. [17] implement data mining techniques inspired by Kurt

Vonnegut's theory of the shape of stories. Vonnegut suggested graphing fictional works based on the fortune of the main character's experiences over the passage of time in the story. Using text sentiment analysis, Reagan et al. scored the emotional content over the course of a novel based on the occurrence of select words in the labMT data set for 1,737 books from the Project Gutenberg database. They found the majority of emotional arcs resided in six classes. In a study of 60 novels, including Jane Austin's *Pride and Prejudice*, Dames et al. [10] determined that the type of narrative is a good predictor for social network structure among characters.

In [4], Beveridge and Shan applied network algorithms on the social network they generated from *A Storm of Swords*, the third novel in George R.R. Martin's *A Song of Ice and Fire* series (which is the literary origin of the HBO drama *Game of Thrones*). Metrics such as PageRank, closeness, betweenness centrality, and modularity provided an empirical approach to determine communities and key characters within the network. Work done by Ribeiro et al. [18] focuses on examining structural properties, such as assortivity and transitivity, of communities in the social network of J.R.R. Tolkien's *The Lord of the Rings* (which included that unabridged novel, along with text from *The Hobbit* and *The Silmarillion*). Beyond static networks, Agarwal et al. [2] analyze the dynamic network for *Alice in Wonderland*, defined by the mining of the ten chapters independently of each other. Such analysis may be important in determining characters with low global importance metrics who are significantly important for part of the story. Deviating from the extraction of character networks, Sack [19] provides a social network generation model for narratives through the concept of *structural balance theory* using signed edges between characters.

2   Experimental Design and Methods

The twin goals of our experiments are to highlight some of the complexities present in character networks via their network properties and to determine a possible synthetic model of the character networks.

2.1   Network Properties

We use the Gephi open source software package to extract communities and compute various network statistics from character networks. These analyses are all done on weighted, undirected, graphs. For community analysis, we use modularity and the Louvain method. Centrality measures are a classic tool in social network analysis to determine the important individuals. They have been found to also serve the same role in character networks. We consider weighted degree, closeness, betweenness, eigencentrality, and PageRank centrality. We briefly review these methods; see [5] for more background on complex network properties. The *closeness* of a node u is the average distance between u and all other nodes (here distance is the standard shortest path metric in graph theory). The *betweenness* of u is the proportion of shortest paths that transit through an u

as an intermediate node. The *eigencentrality* of u is its corresponding coordinate in the largest eigenvector of the weighted adjacency matrix. PageRank centrality is based on the stationary distribution of a random walk on the network that periodically teleports to a node chosen uniformly at random.

2.2   Model Selection

The goal of our model selection experiments is to determine a random graph model that matches empirically observed properties of character networks. Our methodology is to create a compact summary of the network statistics that is invariant to the labeling of the nodes of the network. In other words, we would derive the same statistics if we permuted the adjacency matrix. The summaries we use are the 3-profile, 4-profile, and eigenvalue histogram. The *k-profile* of a graph G counts the number of times each graph on k nodes appears as an induced subgraph of G; see [8]. An *eigenvalue histogram* is a histogram of the eigenvalues of the normalized Laplacian matrix, which all lie between 0 and 2, with equally spaced bins. These techniques are well established in model selection for various types of biological and social networks [6, 14].

In contrast to the previous section we use undirected, unweighted graphs for this experiment. This choice reflects our goal to model the connectivity of the networks, rather than their joint connectivity and weight structure.

We use the algorithm in [9] to compute a global graph 4-profile for each character network. This is a generalization of graphlets [16, 21], a similar method of motif counting for connected subgraphs. One difference is that the 4-profile includes disconnected graphs as well. We use standard algorithms for computing all eigenvalues of the normalized Laplacian where we treat the normalized Laplacian as a dense matrix. We compute a histogram based on five equally spaced bins.

We examine the following random graph models on n nodes, with parameters chosen to match those of the original character network:

1. *Preferential Attachment (PA).* In the PA model, at each step, a node is added to the graph and m edges are placed from the new node to existing nodes. These edges are chosen with probability proportional to the degree of each node before the new node arrived. If m is chosen such that

$$\frac{2}{n} + 2m = \frac{2|E|}{n},$$

   then the number of edges will match that of the original graph in expectation.
2. The *Binomial Random Graph* $G(n, p)$, or *Erdős-Rényi (ER)* model. Each of the $\binom{n}{2}$ edges is connected according to an independent binomial random variable with probability p proportional to the expected average degree. We use $p = |E| / \binom{n}{2}$ to match the average degree of the original network.
3. The *Chung-Lu (CL)* model. The CL model generalizes the binomial random graph model to non-uniform edge probabilities. Graphs in this model are parameterized by an expected degree distribution (the character network's true

degree distribution) rather than a scalar average degree. Each edge is connected with probability proportional to the product of the expected degrees w_i of its endpoints:

$$p_{ij} = \frac{1}{C} w_i w_j.$$

4. The *Configuration Model (CFG)*. In the CFG model, we select a graph uniformly from the set of graphs which exactly match the target degree distribution. In practice, the degree distribution may vary slightly from the target since we disregard self loops and multi-edges created during this process.

Our method to determine the best random graph model fitting the data is to generate samples and train a machine learning algorithm to identify each model. We then ask the algorithm to classify the real graph. First, 100 random graphs from each model are used to train a machine learning classifier. Then in the test step, the classifier predicts a class label for the original character network. This provides a measure of which random graph model best fits the character network. We study the following machine learning algorithms: two variants of linear classifiers (SVMs) and two ensemble methods based on decision trees (Random Forests and Boosted Decision Trees). For more about these models, see [13].

1. *Support Vector Machines (SVM)*. The SVM algorithm is a simple way to classify points in Euclidean space. Geometrically, the binary SVM classifier is defined by a hyperplane **w** that maximally separates points from both classes on either side. This problem can be formulated as a quadratic program with either ℓ_1 or ℓ_2 regularization. Since our application involves more than two classes, a "one-versus-the-rest" classifier is trained for each random graph model. Then we select the model corresponding to the highest confidence score during classification.
2. *Random Forest*. In this algorithm, classifiers combine many weaker decision trees, each working on a random subset of the feature space, to reduce variance and increase robustness. The output is simply a sum of the scores given by each tree.
3. *Boosted Decision Trees*. This algorithm gives another approach to combine several weak learners. We use a popular boosting algorithm called AdaBoost [11] in which new trees are constructed sequentially to correct mistakes made by the previous trees. As before, the final prediction is decided by summing across trees.

2.3   Data

Novels: Our method for extracting character networks from novels begins with the tokenization of an input of text. Character names and aliases are then gathered by the parser, coupled with manual addition and subtraction as needed. Names and aliases representing one character are assigned to its main name. The main names represent the nodes in the network. The parser runs through the text recording the occurrence of two names within a certain number of words apart. For our results, we set the distance parameter to 15 words apart. In the instance where two names share the same keystone and are both within the specified distance along with another name, the parser will record one occurrence between the unique key names. The number of occurrences between two key names represents the weighted edge between the corresponding nodes in the character graph. The node and adjacency lists are recorded via two separate CSV files, which are imported to Gephi, an open source software platform for network analysis and visualization.

The following books were selected for the experiment: *Twilight* by Stephanie Meyer, *Harry Potter and the Goblet of Fire* by J.K. Rowlings, and *The Stand* by Stephen King. We summarize basic network statistics for the novels in Table 1. The results support the view of character networks as complex networks that are dense and small world.

Table 1. Global metrics of character networks from the novels.

Novel	# Nodes	Avg. degree	Avg. weighted degree	Diameter	Edge density	Avg. distance	Clust. coeff.
The Stand	39	14.36	335.33	3	0.378	1.66	0.718
Goblet	62	18.55	305.29	2	0.304	1.69	0.746
Twilight	27	9.11	76.37	4	0.35	1.74	0.783

Moviegalaxies: The website http://moviegalaxies.com/ has assembled a large number of character networks based on movie scripts. There are over 800 networks available. Each network is weighted, although we discard the weights as we only use this for the model selection problem. Some of the properties of these networks are shown in Fig. 1.

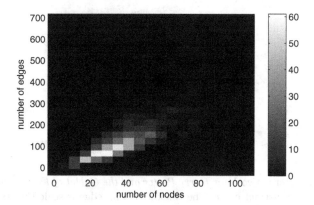

Fig. 1. Number of characters versus number of edges in the *Moviegalaxies* network data. The color shows how many graphs (according to the color bar) lie at the same (nodes, edges) bin.

3   Results

3.1   Analysis of Novel Character Networks

Main characters from each of the novels analyzed scored consistently high in each of the six centrality measures. We present the centrality measures for the top twelve characters from the novel character networks in the figures below. Characters are ranked by increasing PageRank. For example, Harry, Ron and Hermione are identified as the top characters in *Harry Potter and the Goblet of Fire*. Further, our methods accurately predict the community structure for each of the three novels. Visualizations of the character networks and their community in the novels is found below.

For *Harry Potter and the Goblet of Fire* the communities were: Hogwarts, the Dursleys, the Weasleys, Sytherin, and the inseparable friends Seamus and Dean. See Fig. 2.

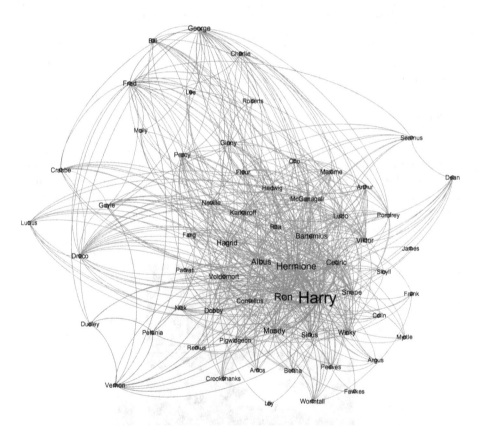

Fig. 2. The character network for *Harry Potter and the Goblet of Fire*. Each community is represented by a distinct color. The thickness of an edge is scaled to its weight, and the size of a name is scaled to the Pagerank score. (Color figure online)

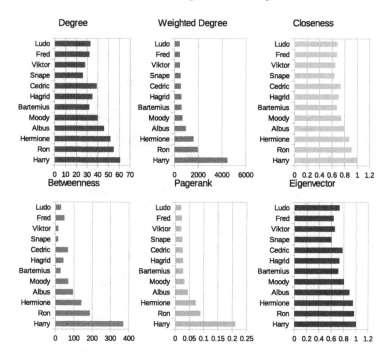

Fig. 3. Centrality measures for *Harry Potter and the Goblet of Fire.*

For *Twilight*, the three communities can be labeled as: vampires, high school students, and characters close to Charlie. See Fig. 4.

For *The Stand*, the government and the evil Las Vegas group emerged as separate communities. The *free zone society* was divided into three groups based on their relation to the main characters, Stu, Larry, and Nick. See Fig. 6.

3.2   Model Selection Results

Hyperparameters for each classifier were selected using stratified, 5-fold cross validation. All features were normalized to have zero mean and unit variance before training. First, the random graph data was split into half training and half holdout. Classification performance on the holdout set verified both the choice of hyperparameters and the separability of classes in our chosen feature space. All classifiers achieved nearly perfect classification on the holdout set, with over 0.98 precision and recall in nearly all cases (often exactly 1). The F1 score was at least 0.97. Thus, our four random graph models represent distinct classes.

Table 2 shows model selection scores for the setup described in Sect. 2.2 (we train on the entire random graph data, and test on the original character network). These scores were calculated differently depending on the classifier. For the SVM algorithms, distance to the separating hyperplane was used. For AdaBoost, we use the final decision function, and soft decision probabilities were

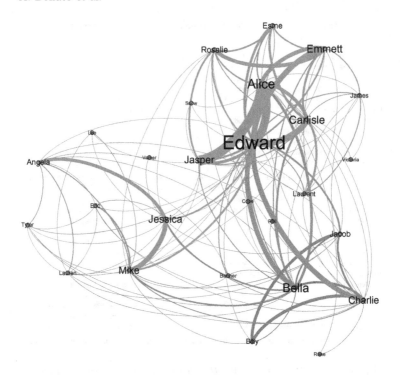

Fig. 4. The character network for *Twilight*.

used for the random forests; see Fig. 8. For all classifiers, a more positive (negative) score indicates more confidence the original graph does (not) belong to the model. Clearly, CL is the best random graph model for all three novels, with each remaining model taking a distant second place in at least one case.

Figure 9 shows our naming convention for the motifs used in our graph profile features. The most important features for the CL SVM hyperplanes were predominantly cliques: induced subgraphs H_3, F_5, F_9, and F_{10}. For the tree-based classifiers, the most important motifs for distinguishing among graph models include some disconnected subgraphs: H_0, H_2, F_2, F_5, and F_{10}. The eigenvalue histograms generally had low importance for all machine learning classifiers. Thus, similar results were obtained using only graph profile features. See Table 3. Figure 10 shows similar aggregate results for the 800 character networks in the *Moviegalaxies* data set, with CL as the best random graph model for the overwhelming number of character networks.

4   Discussion and Future Work

We presented a comparative and quantitative analysis of character networks arising from various novels and films. In particular, we analyzed the weighted social networks from the novels *Twilight The Stand*, and *Harry Potter and the Goblet of*

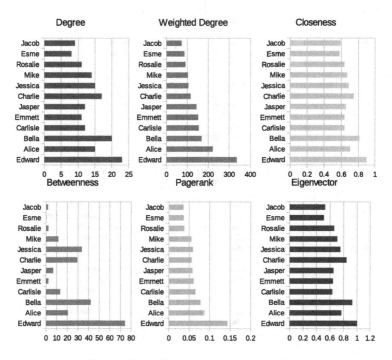

Fig. 5. Centrality measures for *Twilight*.

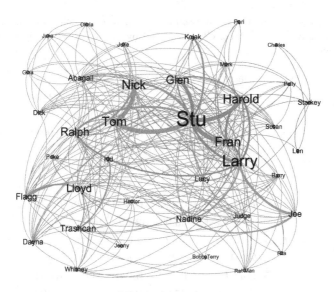

Fig. 6. The character network for *The Stand*.

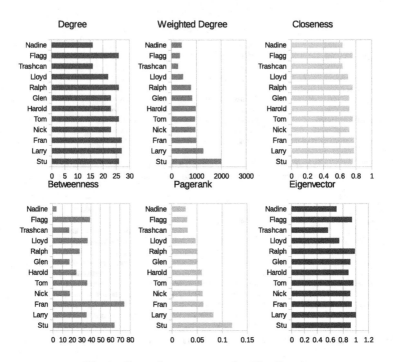

Fig. 7. Centrality measures for *The Stand*.

Table 2. Model selection scores for random graph models using graph profiles and eigenvalue histograms as features. CL is selected by all machine learning classifiers as the best model.

Novel	Classifier	PA	CL	ER	CFG
Goblet	SVM-ℓ_2	2.78	**4.59**	−1.10	−10.65
	SVM-ℓ_1	−0.66	**3.81**	−1.55	−10.80
	Forest	0.00	**0.91**	0.094	0.0011
	AdaBoost	−47.2	**47.4**	25.5	−25.7
Twilight	SVM-ℓ_2	−0.671	**4.49**	−2.98	−9.39
	SVM-ℓ_1	−3.08	**5.19**	−2.06	−12.21
	Forest	0.00083	**0.800**	0.0248	0.175
	AdaBoost	−43.06	**32.30**	10.74	0.0205
The Stand	SVM-ℓ_2	−1.52	**2.65**	−1.24	−3.87
	SVM-ℓ_1	−2.32	**2.87**	−1.14	−4.97
	Forest	0.00	**0.946**	0.00	0.0544
	AdaBoost	−47.04	**50.03**	37.83	−40.82

Table 3. Model selection scores for random graph models using graph profiles alone as features. Once again, CL is selected by all machine learning classifiers as the best model.

Novel	Classifier	PA	CL	ER	CFG
Goblet	SVM-ℓ_2	3.18	**4.44**	−1.15	−10.64
	SVM-ℓ_1	−0.68	**3.81**	−1.53	−10.81
	Forest	0.000	**0.998**	0.002	0.000
	AdaBoost	−47.2	**47.4**	25.5	−25.7
Twilight	SVM-ℓ_2	−0.54	**5.51**	−2.73	−9.52
	SVM-ℓ_1	−2.78	**5.25**	−2.02	−12.24
	Forest	0.00	**1.00**	0.00	0.00
	AdaBoost	−39.72	**34.51**	−7.44	12.66
The Stand	SVM-ℓ_2	−1.18	**2.58**	−1.33	−4.02
	SVM-ℓ_1	−2.35	**2.86**	−1.14	−4.99
	Forest	0.00	**0.94**	0.00	0.06
	AdaBoost	−46.49	**50.32**	38.36	−42.19

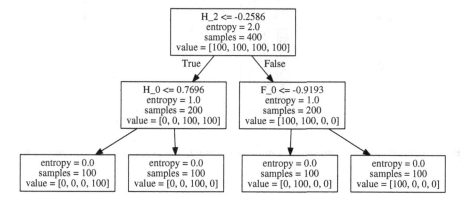

Fig. 8. Example decision tree for the *Goblet* graph.

Fire, along with social networks from 800 films catalogued by [15]. For each of the character networks from the three novels, we extracted the social network from co-occurrence of character names. Community structure was extracted for each network, and statistics such as PageRank and various centrality measures were computed for the characters. In each case, our methodology extracts accurate literary conclusions from the data sets, and successfully identifies the influential characters and the constellations of lesser characters in the books. As pointed out first in [4], the analysis provided of these texts was done algorithmically, without resort to conventional literary analysis.

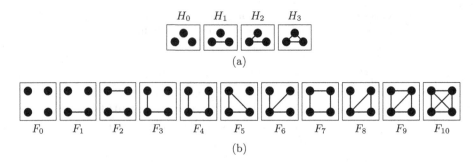

Fig. 9. (a) The four non-isomorphic graphs on 3 nodes that comprise the graph 3-profile. (b) The eleven non-isomorphic graphs on 4 nodes that comprise the graph 4-profile.

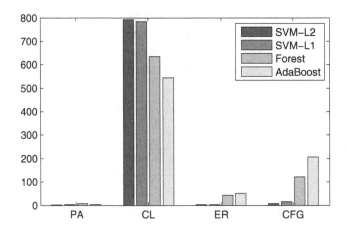

Fig. 10. Summary of *Moviegalaxies* model selection using graph profiles and eigenvalue histograms as features.

For both the novel and http://moviegalaxies.com/ data sets, we employed machine learning techniques to compare and contrast the models against simulated data from popular complex network models. The models considered were the Chung-Lu (CL) model, the configuration model, the PA model, and binomial random graphs. Our methodology used small subgraph counts or motifs as classifiers for the Support Vector Machine (SVM) and other machine learning algorithms. For all the data sets, SVM and the other algorithms clearly separated the models, and indicated that the CL model provided the best alignment with the data.

There are various explanations for the conclusions derived from the model selection experiments. As the character networks we consider have relatively few nodes, they are less likely to exhibit various properties such as power law degree distributions or dimensionality found in various on-line social networks such as Facebook. Hence, preferential attachment (an early and successful adopted model for complex networks) or geometric models may be less relevant for character networks. The CL model has a number of properties amenable to modeling character networks. From a literary perspective, an author may intuit a hierarchy of character influence (separated by the degrees of the nodes representing characters), then randomly generate the social ties in the fictional work to complete the network. For instance, Rowlings may have decided in the Harry Potter series that the main triad was Harry, Hermione and Ron, and then gradually added lesser characters revolving around this triad. In terms of the various models, the CL model has 4-node subgraph counts that more accurately model character networks. This is likely due to the property of CL graphs that they have a more diverse set of dense subgraph structures that are more closely related to those that appear in character networks. We plan to continue investigating this finding that CL graphs are good matches for character networks.

In future work, we plan on expanding our analysis of literary works using Project Gutenberg and other sources. We will also explore other models such as random geometric graphs and Kronecker graphs. More broadly, our approach and those of other recent works [2,4,17,18], represents a trend towards the algorithmic and big data-theoretic analysis of cultural works. Such a direction may lead to new models for the evolution and construction of character networks, and a broader view of such networks as complex and evolving.

References

1. Alberich, R., Miro-Julia, J., Rossello, F.: Marvel Universe looks almost like a real social network. arXiv preprint arXiv:0202174 (2002)
2. Agarwal, A., Corvalan, A., Jensen, J., Rambow, O.: Social network analysis of "Alice in Wonderland". In: Proceedings of the Workshop on Computational Linguistics for Literature (2012)
3. Bastian, M., Heymann, S., Jacomy, M.: Gephi: an open source software for exploring and manipulating networks, In: Proceedings of the International AAAI Conference on Weblogs and Social Media (2009)
4. Beveridge, A., Shan, J.: Network of thrones. Math Horizons Mag. **23**, 18–22 (2016)
5. Bonato, A.: A Course on the Web Graph. American Mathematical Society Graduate Studies Series in Mathematics, Providence (2008)
6. Bonato, A., Gleich, D.F., Kim, M., Mitsche, D., Pralat, P., Tian, A., Young, S.J.: Dimensionality matching of social networks using motifs and eigenvalues. PLOS ONE **9**(9), e106052 (2014)
7. Bonato, A., Tian, A.: Complex networks and social networks. In: Kranakis, E. (ed.) Social Networks. Mathematics in Industry series. Springer, Heidelberg (2011)
8. Elenberg, E.R., Shanmugam, K., Borokhovich, M., Dimakis, A.G.: Beyond triangles: a distributed framework for estimating 3-profiles of large graphs, In: Proceedings of ACM SIGKDD Conference on Knowledge Discovery and Data Mining (2015)
9. Elenberg, E.R., Shanmugam, K., Borokhovich, M., Dimakis, A.G.: Distributed estimation of graph 4-profiles. In: Proceedings of International World Wide Web Conference (2016)

10. Elson, D., Dames, N., McKeown, K.: Extracting social networks from literary fiction. In: Proceedings of the 48th Annual Meeting of the Association for Computational Linguistics (2010)
11. Freund, Y., Schapire, R.: A decision-theoretic generalization of on-line learning and an application to boosting. J. Comput. Syst. Sci. **55**, 119–139 (1995)
12. Gleiser, P.M.: How to become a superhero. J. Stat. Mech.:Theor. Exp. **2007**, P09020 (2007)
13. Hastie, T., Tibshirani, R., Friedman, J.: The Elements of Statistical Learning. Springer, Heidelberg (2009)
14. Janssen, J., Hurshman, M., Kalyaniwalla, N.: Model selection for social networks using graphlets. Internet Math. **8**, 338–363 (2012)
15. Kaminski, J., Schober, M., Albaladejo, R., Zastupailo, O., Hidalgo, C.: Moviegalaxies - social networks in movies (2011). http://moviegalaxies.com. Accessed September 2016
16. Pržulj, N.: Biological network comparison using graphlet degree distribution. Bioinformatics **23**, 77–183 (2007)
17. Reagan, A.J., Mitchell, L., Kiley, D., Danforth, C.M., Dodds, P.S.: The emotional arcs of stories are dominated by six basic shapes, arXiv preprint arXiv:1606.07772 (2016)
18. Ribeiro, M.A., Vosgerau, R.A., Andruchiw, M.L.P., Ely de Souza Pinto, S.: The complex social network from the Lord of the Rings. Rev. Bras. Ensino Fs **38**, 1304 (2016)
19. Sack, G.: Character networks for narrative generation. In: Intelligent Narrative Technologies: Papers from the AIIDE Workshop, AAAI Technical Report WS-12-14 (2012)
20. Shalev-Shwartz, S., Ben-David, S.: Understanding Machine Learning from Theory to Algorithms. Cambridge University Press, Cambridge (2014)
21. Shervashidze, N., Vishwanathan, S.V.N., Petri, T.H., Mehlhorn, K., Borgwardt, K.M.: Efficient graphlet kernels for large graph comparison, In: Proceedings of the 20th International Conference on Artificial Intelligence and Statistics (2009)
22. West, D.B.: Introduction to Graph Theory, 2nd edn. Prentice Hall, Upper Saddle River (2001)

Modularity of Complex Networks Models

Liudmila Ostroumova Prokhorenkova[1,2]([✉]),
Paweł Prałat[3,4], and Andrei Raigorodskii[1,2,5,6]

[1] Moscow Institute of Physics and Technology, Moscow, Russia
ostroumova-la@yandex.ru
[2] Yandex, Moscow, Russia
[3] Ryerson University, Toronto, ON, Canada
[4] The Fields Institute for Research in Mathematical Sciences, Toronto, ON, Canada
[5] Moscow State University, Moscow, Russia
[6] Buryat State University, Ulan-ude, Buryat Republic, Russia

Abstract. Modularity is designed to measure the strength of division of a network into clusters (known also as communities). Networks with high modularity have dense connections between the vertices within clusters but sparse connections between vertices of different clusters. As a result, modularity is often used in optimization methods for detecting community structure in networks, and so it is an important graph parameter from practical point of view. Unfortunately, many existing non-spatial models of complex networks do not generate graphs with high modularity; on the other hand, spatial models naturally create clusters. We investigate this phenomenon by considering a few examples from both sub-classes. We prove precise theoretical results for the classical model of random d-regular graphs as well as the preferential attachment model, and contrast these results with the ones for the spatial preferential attachment (SPA) model that is a model for complex networks in which vertices are embedded in a metric space, and each vertex has a sphere of influence whose size increases if the vertex gains an in-link, and otherwise decreases with time.

1 Introduction and Definitions

Many social, biological, and information systems can be represented by networks, whose vertices are items and links are relations between these items [2,4,6,12]. That is why the evolution of complex networks attracted a lot of attention in recent years and there has been a great deal of interest in modeling of these networks [9,17,30]. The hyperlinked structure of the Web, citation patterns, friendship relationships, infectious disease spread, these are seemingly disparate linked data sets which have fundamentally very similar natures. Indeed, it turns out that many real-world networks have some typical properties: power-law degree distribution, small diameter, high clustering coefficient, and others [27,29,33]. Such properties are well-studied both in real-world networks and in many theoretical models.

© Springer International Publishing AG 2016
A. Bonato et al. (Eds.): WAW 2016, LNCS 10088, pp. 115–126, 2016.
DOI: 10.1007/978-3-319-49787-7_10

Another important property of complex networks is their community structure, that is, the organization of vertices in clusters, with many edges joining vertices of the same cluster and comparatively few edges joining vertices of different clusters [14,18]. In social networks communities may represent groups by interest, in citation networks they correspond to related papers, in the Web communities are formed by pages on related topics, etc. Being able to identify communities in a network could help us to exploit this network more effectively. For example, clusters in citation graphs may help to find similar scientific papers, discovering users with similar interests is important for targeted advertisement, clustering can also be used for network compression and visualization.

The key ingredient for many clustering algorithms is *modularity*, which is at the same time a global criterion to define communities, a quality function of community detection algorithms, and a way to measure the presence of community structure in a network. Modularity was introduced by Newman and Girvan [31] and it is based on the comparison between the actual density of edges inside a community and the density one would expect to have if the vertices of the graph were attached at random, regardless of community structure.

Unfortunately, modularity is not a well studied parameter for the existing random graph models, at least from a rigorous, theoretical point of view. We are only aware about results for binomial random graphs $G(n,p)$ and random d-regular graphs (see Sect. 2.3 for more details). In this paper, we continue investigating random d-regular graphs and obtain new upper bounds for their modularity. Then we move to the *preferential attachment model*, introduced by Barabási and Albert [5], which is probably the most well-studied model of complex networks. For this model no results on modularity are known and we obtain some preliminary results, both lower and upper bounds, and will investigate this model more in the journal version of this paper. In fact, the lower bound we present holds for all graphs with average degree d and sublinear maximum degree.

As expected, the models discussed above, as well as many others, have a common weakness of low modularity. One family of models which overcomes this deficiency is the family of spatial (or geometric) models, wherein the vertices are embedded in a metric space such that similar vertices are closer to each other than dissimilar ones. The underlying geometry of spatial models naturally leads to the emergence of clusters. We prove this statement rigorously for one example of a geometric model, the Spatial Preferential Attachment model introduced in [1].

The paper is structured as follows. In the next section, we formally define modularity, discuss several random graph models and present known results on modularity in these models. In Sects. 3, 4 and 5 we analyze modularity in random d-regular graphs, preferential attachment and SPA models, respectively.

Due to the space limitations, proofs of our results are omitted in this short proceeding version but will be included in the longer journal one.

2   Preliminaries

2.1   Modularity

The definition of modularity was first introduced by Newman and Girvan in [31]. Since then, many popular and applied algorithms used to find clusters in large data-sets are based on finding partitions with high modularity [16,22,28]. The modularity function favors partitions in which a large proportion of the edges falls entirely within the parts and biases against having too few or too unequally sized parts. Formally, for a given partition $\mathcal{A} = \{A_1, \ldots, A_k\}$ of the vertex set $V(G)$, let

$$q_{\mathcal{A}} = \sum_{A \in \mathcal{A}} \left(\frac{e(A)}{|E(G)|} - \frac{(\sum_{v \in A} \deg(v))^2}{4|E(G)|^2} \right), \tag{1}$$

where $e(A) = |\{uv \in E(G) : u, v \in A\}|$ is the number of edges in the graph induced by the set A. The first term, $\sum_{A \in \mathcal{A}} \frac{e(A)}{|E(G)|}$, is called the *edge contribution*, whereas the second one, $\sum_{A \in \mathcal{A}} \frac{(\sum_{v \in A} \deg(v))^2}{4|E(G)|^2}$, is called the *degree tax*. It is easy to see that $q_{\mathcal{A}}$ is always smaller than one. Also, if $\mathcal{A} = \{V(G)\}$, then $q_{\mathcal{A}} = 0$.

The *modularity* $q^*(G)$ is defined as the maximum of $q_{\mathcal{A}}$ over all possible partitions $\mathcal{A}$ of $V(G)$; that is,

$$q^*(G) = \max_{\mathcal{A}} q_{\mathcal{A}}(G).$$

In order to maximize $q_{\mathcal{A}}(G)$ one wants to find a partition with large edge contribution subject to small degree tax. If $q^*(G)$ approaches 1 (which is the maximum possible value), we observe a strong community structure; conversely, if $q^*(G)$ is close to zero, we are given a graph with no community structure.

2.2   Random Graph Models

Classical Models. Let us recall two classical models of random graphs that are extensively studied in the literature. The *binomial random graph* $\mathcal{G}(n, p)$ is the random graph G with the vertex set $[n] := \{1, 2, \ldots, n\}$ in which every pair $\{i, j\} \in \binom{[n]}{2}$ appears independently as an edge in G with probability p. Note that $p = p(n)$ may (and usually does) tend to zero as n tends to infinity.

However, in this paper we concentrate on another probability space, the probability space of *random d-regular graphs* with uniform probability distribution. This space is denoted $\mathcal{G}_{n,d}$, and asymptotics are for $n \to \infty$ with $d \geq 2$ fixed, and n even if d is odd.

We say that an event in a probability space holds *asymptotically almost surely* (or *a.a.s.*) if the probability that it holds tends to 1 as n goes to infinity. Since we aim for results that hold a.a.s., we will always assume that n is large enough.

Preferential Attachment. The *Preferential Attachment* (PA) model, introduced by Barabási and Albert [5], was an early stochastic model of complex networks. We will use the following precise definition of the model, as considered by Bollobás and Riordan in [10] as well as Bollobás et al. [11].

Let G_1^0 be the null graph with no vertices (or let G_1^1 be the graph with one vertex, v_1, and one loop). The random graph process $(G_1^t)_{t \geq 0}$ is defined inductively as follows. Given G_1^{t-1}, we form G_1^t by adding a vertex v_t together with a single edge between v_t and v_i, where i is selected randomly with the following probability distribution:

$$\mathbb{P}(i = s) = \begin{cases} \deg(v_s, t-1)/(2t-1) & 1 \leq s \leq t-1, \\ 1/(2t-1) & s = t, \end{cases}$$

where $\deg(v_s, t-1)$ denotes the degree of v_s in G_1^{t-1}. In other words, we send an edge e from v_t to a random vertex v_i, where the probability that a vertex is chosen is proportional to its degree at the time, counting e as already contributing one to the degree of v_t.

For $m \in \mathbb{N} \setminus \{1\}$, the process $(G_m^t)_{t \geq 0}$ is defined similarly with the only difference that m edges are added to G_m^{t-1} to form G_m^t (one at a time), counting previous edges as already contributing to the degree distribution. Equivalently, one can define the process $(G_m^t)_{t \geq 0}$ by considering the process $(G_1^t)_{t \geq 0}$ on a sequence $v_1', v_2', \ldots$ of vertices; the graph G_m^t is formed from G_1^{tm} by identifying vertices $v_1', v_2', \ldots, v_m'$ to form v_1, identifying vertices $v_{m+1}', v_{m+2}', \ldots, v_{2m}'$ to form v_2, and so on. Note that in this model G_m^t is in general a multigraph, possibly with multiple edges between two vertices (if $m \geq 2$) and self-loops.

It was shown in [11] that for any $m \in \mathbb{N}$ a.a.s. the degree distribution of G_m^n follows a power law: the number of vertices with degree at least k falls off as $(1 + o(1))ck^{-2}n$ for some explicit constant $c = c(m)$ and large $k \leq n^{1/15}$. Also, in the case $m = 1$, G_1^n is a forest. Each vertex sends an edge either to itself or to an earlier vertex, so the graph consists of components which are trees, each with a loop attached. The expected number of components is then $\sum_{t=1}^n 1/(2t-1) \sim (1/2) \log n$ and, since events are independent, we derive that a.a.s. there are $(1/2 + o(1)) \log n$ components in G_1^n by Chernoff's bound. Moreover, Pittel [32] essentially showed that a.a.s. the largest distance between two vertices in the same component of G_1^n is $(\gamma^{-1} + o(1)) \log n$, where γ is the solution of $\gamma e^{1+\gamma} = 1$. In contrast, for the case $m \geq 2$ it is known that a.a.s. G_m^n is connected and its diameter is $(1 + o(1)) \log n / \log \log n$ [10].

Spatial Preferential Attachment. The *Spatial Preferential Attachment* (SPA) model [1], designed as a model for the World Wide Web, combines geometry and preferential attachment, as its name suggests. Setting the SPA model apart is the incorporation of 'spheres of influence' to accomplish preferential attachment: the greater the degree of a vertex, the larger its sphere of influence, and hence the higher the likelihood of the vertex gaining more neighbors.

We now give a precise description of the SPA model. Let $S = [0, 1]^m$ be the unit hypercube in $\mathbb{R}^m$, equipped with the torus metric derived from any of the L_p norms. This means that for any two points x and y in S,

$$d(x,y) = \min \left\{ ||x - y + u||_p : u \in \{-1,0,1\}^m \right\}.$$

The torus metric thus 'wraps around' the boundaries of the unit square; this metric was chosen to eliminate boundary effects. The parameters of the model consist of the *link probability* $p \in [0,1]$, and two positive constants A_1 and A_2, which, in order to avoid the resulting graph becoming too dense, must be chosen so that $pA_1 < 1$. The SPA model generates stochastic sequences of directed graphs $(G_t : t \geq 0)$, where $G_t = (V_t, E_t)$, and $V_t \subseteq S$. Let $\deg^-(v,t)$ be the in-degree of the vertex v in G_t, and $\deg^+(v,t)$ its out-degree. We define the *sphere of influence* $S(v,t)$ of the vertex v at time $t \geq 1$ to be the ball centered at v with volume $|S(v,t)|$ defined as follows:

$$|S(v,t)| = \frac{A_1 \deg^-(v,t) + A_2}{t}, \tag{2}$$

or $S(v,t) = S$ and $|S(v,t)| = 1$ if the right-hand-side of (2) is greater than 1.

The process begins at $t = 0$, with G_0 being the null graph. Time-step t, $t \geq 1$, is defined to be the transition between G_{t-1} and G_t. At the beginning of each time-step t, a new vertex v_t is chosen *uniformly at random* from S, and added to V_{t-1} to create V_t. Next, independently, for each vertex $u \in V_{t-1}$ such that $v_t \in S(u, t-1)$, a directed link (v_t, u) is created with probability p. Thus, the probability that a link (v_t, u) is added in time-step t equals $p\,|S(u, t-1)|$.

The SPA model produces scale-free networks, which exhibit many of the characteristics of real-life networks (see [1,13]). In [19], it was shown that the SPA model gave the best fit, in terms of graph structure, for a series of social networks derived from Facebook. In [20], some properties of common neighbors were used to explore the underlying geometry of the SPA model and quantify vertex similarity based on distance in the space. However, the distribution of vertices in space was assumed to be uniform [20] and so in [21] non-uniform distributions were investigated which is clearly a more realistic setting.

2.3   Previous Results on Modularity

In this section we discuss known bounds for modularity in different random graph models.

The *isoperimetric number* of a graph G is defined as

$$i(G) = \min_{V(G) = V_1 \cup V_2} \frac{e(V_1, V_2)}{\min\{|V_1|, |V_2|\}},$$

where $e(V_1, V_2) = |\{uv \in E(G) : u \in V_1, v \in V_2\}|$ is the number of edges between the sets V_1 and V_2. The following result was shown by McDiarmid and Skerman in [23]. Let G be any d-regular graph on n vertices. As mentioned in [23], the following useful upper bound on the modularity is almost immediate:

$$q^*(G) \leq \max\{1 - i(G)/d, 3/4\}. \tag{3}$$

Turning to random d-regular graphs, Bollobás in [8] showed that a.a.s. $i(\mathcal{G}_{n,d}) \geq (1 - \eta)d/2$, where $0 < \eta < 1$ is such that $2^{4/d} < (1 - \eta)^{1-\eta}(1 + \eta)^{1+\eta}$ and so a.a.s.

$$q^*(\mathcal{G}_{n,d}) \leq U_1 = U_1(d) := \max\{1/2 + \eta/2, 3/4\}.$$

As a result, we get the first non-trivial upper bounds for $q^*(\mathcal{G}_{n,d})$ presented in Table 1 that hold a.a.s.

In [23], the bound (3) was slightly improved when the maximum size of parts in our partition is restricted. Formally, given $\delta > 0$, for a graph G with $n \geq 1/\delta$ vertices, they define $q_\delta(G)$ to be the maximum modularity of all partitions for G such that each part has size at most δn. They show that for any $\varepsilon > 0$ there exists $\delta > 0$ such that any d-regular graph with at least $1/\delta$ vertices satisfies

$$q_\delta(G) \leq 1 - 2i(G)/d + \varepsilon.$$

Again, using the result of Bollobás we get that there exists $\delta > 0$ such that

$$U_2 = U_2(d) := \eta + \varepsilon$$

serves as an upper bound that holds a.a.s. for $q_\delta(\mathcal{G}_{n,d})$; again, see Table 1 for numerical values for small values of d. It is straightforward to see that $i(G) \geq d/2 - \sqrt{(\log 2)d}$ (see, for example, [8]) and so, in particular, U_2 can be made arbitrarily small by taking d large enough (and δ small enough). However, let us note that these upper bounds for $q_\delta(\mathcal{G}_{n,d})$, while useful, cannot be directly translated into any bound for $q^*(\mathcal{G}_{n,d})$.

Table 1. Upper bounds for $q^*(\mathcal{G}_{n,d})$ and for $q_\delta(\mathcal{G}_{n,d})$ (U_2)

d	U_1	U_2	U_3
3	0.9386	0.8771	0.8038
4	0.8900	0.7800	0.6834
5	0.8539	0.7078	0.6024
6	0.8261	0.6521	0.5435
7	0.8038	0.6076	0.4984
8	0.7855	0.5710	0.4624
9	0.7702	0.5403	0.4330
10	0.7570	0.5140	0.4083

Investigating random d-regular graphs continues in [24], a very recent paper. In fact, some of our results for this model mentioned below are obtained independently there. Moreover, they investigate the class of graphs whose product of treewidth and maximum degree is much less than the number of edges. This shows, for example, that random planar graphs typically have modularity close to 1, which is another indication that clusters naturally emerge where geometry is included.

3   Random d-regular Graphs

3.1   Lower Bound

For completeness, let us briefly discuss the following known lower bound for the modularity of $\mathcal{G}_{n,d}$. It is known that a.a.s. for any $d \in \mathbb{N} \setminus \{1, 2\}$, $\mathcal{G}_{n,d}$ is Hamiltonian. As pointed out in [23], one can use this fact to partition the graph such that it breaks the cycle into $\lceil \sqrt{n} \rceil$ paths of length at most $\lceil \sqrt{n} \rceil$. For this particular partition the edge contribution is $2/d - O(1/\sqrt{n})$ and the degree tax is $O(1/\sqrt{n})$. It follows then that a.a.s.

$$q^*(\mathcal{G}_{n,d}) \geq \frac{2}{d} - O(1/\sqrt{n}) = \frac{2 + o(1)}{d}.$$

(Our more general lower bound that holds for graphs with average degree d implies the same—see Theorem 4 for more.) Whereas this trivial lower bound could be sharp for $d = 3$ it is definitely not the case for large d. As pointed out in [24], there exists a universal constant $c > 0$ such that a.a.s. $q^*(\mathcal{G}_{n,d}) \geq c/\sqrt{d}$.

3.2   Slightly Improved, Numerical Upper Bound

Let us consider the following, natural, approach that already improves slightly an upper bound for $q^*(\mathcal{G}_{n,d})$. Consider any d-regular graph with n vertices. For a given partition $\mathcal{A} = \{A_1, \ldots, A_k\}$ of the vertex set $V(G)$, let $x_i = |A_i|/n$ and $y_i = 2|E(A_i)|/|A_i|$; that is, set A_i has $x_i n$ vertices and induces $y_i x_i n/2$ edges. Then (1) can be simplified to

$$q_{\mathcal{A}} = \sum_{i=1}^{k} x_i \left(\frac{y_i}{d} - x_i \right). \tag{4}$$

As it is simply a weighted average, $q_{\mathcal{A}} \geq U$ would imply that there exists some set of size xn that induces $yxn/2$ edges, and $y/d - x \geq U$. Using the pairing model [7], we will show that a.a.s. it is not the case (for some carefully chosen $U = U(d)$) and, as a result, it will yield an upper bound for $q^*(\mathcal{G}_{n,d})$ that holds a.a.s.

For a given $d \in \mathbb{N} \setminus \{1, 2\}$, let

$$f(x, y, d) := x(y/2 - 1) \log(x) + (1 - x)(d - 1) \log(1 - x) + d \log(d)/2 \tag{5}$$
$$- xy \log(y)/2 - x(d - y) \log(d - y) - (d - 2xd + xy) \log(d - 2xd + xy)/2.$$

It will be clear once we establish the connection between the function f and random d-regular graphs, but it is straightforward to see that for any $x \in (0, 1)$ we have $f(x, d, d) < 0$ and $f(x, y, d) > 0$ for some $y \in (0, d)$. Indeed, for example note that for a fixed $x \in (0, 1/2]$, $f(x, y, d)$ is strictly concave in $y \in (0, d)$ as

$$\frac{d^2 f(x, y, d)}{dy^2} = \frac{-(d(1 - 2x) + y)dx}{2(d(1 - 2x) + xy)(d - y)y} < 0.$$

Let $y_3 = y_3(x, d)$ be the largest value of $y \in (0, 1)$ such that $f(x, y, d) = 0$; in particular, $f(x, y, d) \leq 0$ for any $y \in (y_3, d)$. Finally, let

$$U_3 = U_3(d) := \sup_{x \in (0,1)} \left(\frac{y_3(x, d)}{d} - x \right).$$

As usual, see Table 1 for numerical values for small values of d. The promised upper bound follows immediately from the following theorem.

Theorem 1. *Let $d \in \mathbb{N} \setminus \{1, 2\}$ and $\varepsilon > 0$ be an arbitrarily small constant. The following property holds a.a.s. for $\mathcal{G}_{n,d}$. No set A of size xn (for any $x = x(n) \in (0, 1)$) induces a graph with $yxn/2$ edges, where $y_3(x, d) + \varepsilon \leq y \leq d$ and $y_3(x, d)$ is defined as above. In particular, this implies that*

$$q^*(\mathcal{G}_{n,d}) \leq U_3 + \varepsilon/d,$$

where $U_3 = U_3(d)$ is defined as above.

3.3   Explicit but Weaker Upper Bound

Theorem 1 provides an upper bound that can be easily numerically computed for a given $d \in \mathbb{N} \setminus \{1, 2\}$. Next, we present a slightly weaker but an explicit bound that can be obtained using the expansion properties of random d-regular graphs that follow from their eigenvalues. In particular, it will imply that a.a.s. $q^*(\mathcal{G}_{n,d}) = O(1/\sqrt{d})$ and so $q^*(\mathcal{G}_{n,d}) \to 0$ as $d \to \infty$.

Theorem 2. *Let $d \in \mathbb{N} \setminus \{1, 2\}$ and $\varepsilon > 0$ be an arbitrarily small constant. The following property holds a.a.s. for $\mathcal{G}_{n,d}$. No set A of size xn induces a graph with more than $yxn/2$ edges, where $y = dx + \lambda(1 - x)$. In particular, this implies that a.a.s.*

$$q^*(\mathcal{G}_{n,d}) \leq \frac{\lambda}{d} \leq \frac{2\sqrt{d-1} + \varepsilon}{d} \leq \frac{2}{\sqrt{d}}.$$

4   The Preferential Attachment Model

4.1   Constant Average Degree Graphs

In order to obtain a lower bound for modularity of Preferential Attachment graphs, we first analyze graphs with constant average degree in general. In this section, we extend the results of [26] and we start with the analysis of trees. It was proven in [26] that trees with maximum degree $\Delta = o(\sqrt[5]{n})$ have asymptotic modularity 1. We generalize this result in two ways: first, we relax the condition on maximum degree; second, we allow our graphs to be disconnected, that is, we consider forests instead of trees. We prove the following theorem.

Theorem 3. *Let $\{F_n\}$ be a sequence of forests, F_n is a forest on n vertices with no isolated ones and $\Delta = \Delta(F_n) = o(n)$. Then $q^*(F_n) \geq 1 - O\left(\sqrt{\frac{\Delta}{n}}\right) = 1 - o(1)$ as $n \to \infty$.*

Note that it is also known that the asymptotic modularity of trees with maximum degree $\Delta = \Omega(n)$ is strictly less than 1 [26]. Hence, the assumption $\Delta = o(n)$ cannot be eliminated. Now we can use the previous theorem to get the following result for graphs of bounded average degree.

Theorem 4. *Let $\{G_n\}$ be a sequence graphs, G_n is a connected graph on n vertices with the average degree $\frac{2|E(G_n)|}{n} \leq D$ for some constant D, and $\Delta = \Delta(G_n) = o(n)$. Then $q^*(G_n) \geq \frac{2}{D} - O\left(\sqrt{\frac{\Delta}{n}}\right) = \frac{2}{D} - o(1)$.*

4.2 Lower Bound

The following theorem easily follows from the above result.

Theorem 5. *For any $\varepsilon > 0$ a.a.s.*

$$q^*(G_m^n) \geq \frac{1}{m} - O\left(n^{-1/4+\varepsilon}\right) = \frac{1}{m} - o(1).$$

As in the case of random d-regular graphs, it is natural to conjecture that the above lower bound is not sharp. Let $c \in (0,1)$ and consider the following partition: $A_1 = \{v_1, \ldots, v_{cn}\}$, $A_2 = V(G_m^n) \setminus A_1 = \{v_{cn+1}, \ldots, v_n\}$. Using martingales, it is possible to show that a.a.s. $\sum_{v \in A_1} \deg(v,n) \sim 2mn\sqrt{c}$ (and so $\sum_{v \in A_2} \deg(v,n) \sim 2mn(1 - \sqrt{c})$). Clearly, $e(A_1) = mcn$ and so a.a.s. $e(A_1, A_2) \sim 2mn(\sqrt{c} - c)$ and $e(A_2) \sim mn(1 + c - 2\sqrt{c})$. The edge contribution and the degree tax are then both asymptotic to $1 + 2c - 2\sqrt{c}$. Not surprisingly, such partition cannot be used to get a non-trivial lower bound for the modularity but, similarly to the situation for random d-regular graphs, we may try to use it as a starting point to get slightly better partition. The basic idea is very simple: one can start with a given partition (or partition the vertices randomly into two classes), and if a vertex has more neighbors in the other class than in its own, then we randomly decide whether to shift it to the other class or leave it where it is. This approach proved to be useful to get a bound for the bisection width in random d-regular graphs [3] which, in turn, yields a lower bound for the modularity [24]. We plan to investigate it further in the journal version of this paper.

4.3 Upper Bound

The *edge expansion* $\rho = \rho(G)$ of a graph G is defined as follows:

$$\rho = \min_{S \subset V(G), |S| \leq |V|/2} \frac{e(S, V \setminus S)}{|S|}.$$

In [25] it was shown that a.a.s. $\rho(G_m^n) \geq \alpha$, provided that $2(m-1) - 4\alpha - 1 > 0$. In other words, for any $\varepsilon > 0$ we have that a.a.s.

$$\rho(G_m^n) \geq \frac{m}{2} - \frac{3+\varepsilon}{4}.$$

Using this observation one can easily obtain a non-trivial upper bound for $q^*(G_m^n)$.

Let $\varepsilon > 0$ be an arbitrary small constant. Consider any partition $\mathcal{A} = \{A_1, \ldots, A_k\}$ of the vertex set $V(G_m^n)$. If $|A_i| > n/2$ for some i, then the degree tax is at least

$$\frac{(\sum_{v \in A_i} \deg(v,n))^2}{4|E(G_m^n)|} \geq \frac{(m|A_i|)^2}{4(mn)^2} = \frac{1}{16}.$$

On the other hand, if $|A_i| \leq n/2$ for all i, then a.a.s. the number of edges between parts is equal to

$$\frac{1}{2} \sum_{i=1}^{k} e(A_i, V \setminus A_i) \geq \frac{1}{2} \sum_{i=1}^{k} \rho |A_i| = \frac{\rho n}{2} \geq \left(\frac{m}{4} - \frac{3+\varepsilon}{8} \right) n,$$

and so the edge contribution is a.a.s. at most

$$1 - \left(\frac{1}{4} - \frac{3+\varepsilon}{8m} \right) = \frac{3}{4} + \frac{3+\varepsilon}{8m} \leq \frac{15+\varepsilon}{16},$$

for any $m \geq 2$. The following result holds.

Theorem 6. *For any $\varepsilon > 0$ a.a.s.*

$$q^*(G_2^n) \leq \frac{15+\varepsilon}{16}.$$

Moreover, for any $m \geq 3$ a.a.s.

$$q^*(G_m^n) \leq \frac{15}{16}.$$

Much stronger expansion property was recently obtained in [15]. We are currently working on using this property to obtain general upper bound for $q^*(G_m^n)$ that holds for any integer m as well as specific stronger bounds for small values of m. Details will be provided in the journal version of this paper.

5   The Spatial Preferential Attachment Model

Consider $G_n = (V_n, E_n)$, a graph generated by the SPA model. As the modularity is defined for undirected graphs, we consider $\hat{G}_n$ that is a graph obtained from G_n by replacing each directed edge (u, v) by undirected edge uv. (As edges in G_n are always from 'younger' to 'older' vertices, there is no problem with generating multigraph; $\hat{G}_n$ is a simple graph.) Let us recall that $V_n \subseteq S$ where S is the unit hypercube $[0, 1]^m$. We use the geometry of the model to obtain a suitable partition that yields high modularity of G_n. The following properties (proved many times; see, for example, [1,13]) are the only properties of the model that are used in the proof: a.a.s. for every pair i, t such that $1 \leq i \leq t \leq n$ we have that

$$\deg^-(v_i, t) = O\Big((t/i)^{pA_1} \log^2 n\Big), \tag{6}$$

$$\deg^+(v_i, t) = O\Big(\log^2 n\Big), \tag{7}$$

and $|E(G_n)| = \Theta(n)$. Now, we are ready to state our result for the SPA model.

Theorem 7. *Let $p \in (0, 1]$, $A_1, A_2 > 0$, and suppose that $pA_1 < 1$. Then, the following holds a.a.s.:*

$$q^*(\hat{G}_n) = 1 - O\left(n^{\max\{-1/m, -1+pA_1\}/2} \log^{9/2} n\right) = 1 - o(1).$$

Acknowledgements. This work is supported by Russian Science Foundation (grant number 16-11-10014), NSERC, The Tutte Institute for Mathematics and Computing, and Ryerson University.

References

1. Aiello, W., Bonato, A., Cooper, C., Janssen, J., Prałat, P.: A spatial web graph model with local influence regions. Internet Math. **5**, 175–196 (2009)
2. Albert, R., Barabási, A.-L.: Statistical mechanics of complex networks. Rev. Modern Phys. **74**, 47–97 (2002)
3. Alon, N.: On the edge-expansion of graphs. Comb. Prob. Comput. **6**, 145–152 (1997)
4. Bansal, S., Khandelwal, S., Meyers, L.A.: Exploring biological network structure with clustered random networks. BMC Bioinform. **10**, 405 (2009)
5. Barabási, A.L., Albert, R.: Emergence of scaling in random networks. Science **286**, 509–512 (1999)
6. Boccaletti, S., Latora, V., Moreno, Y., Chavez, M., Hwang, D.-U.: Complex networks: structure and dynamics. Phys. Rep. **424**(45), 175–308 (2006)
7. Bollobás, B.: A probabilistic proof of an asymptotic formula for the number of labelled regular graphs. Eur. J. Combin. **1**(4), 311–316 (1980)
8. Bollobás, B.: The isoperimetric number of random regular graphs. Eur. J. Comb. **9**, 241–244 (1988)
9. Bollobás, B., Riordan, O.M.: Mathematical results on scale-free random graphs. In: From the Genome to the Internet, Handbook of Graphs and Networks, pp. 1–34 (2003)
10. Bollobás, B., Riordan, O.: The diameter of a scale-free random graph. Combinatorica **24**, 5–34 (2004)
11. Bollobás, B., Riordan, O., Spencer, J., Tusnády, G.: The degree sequence of a scale-free random graph process. Random Struct. Algorithms **18**, 279–290 (2001)
12. Borgs, C., Brautbar, M., Chayes, J., Khanna, S., Lucier, B.: The power of local information in social networks. In: Goldberg, P.W. (ed.) WINE 2012. LNCS, vol. 7695, pp. 406–419. Springer, Heidelberg (2012). doi:10.1007/978-3-642-35311-6_30
13. Cooper, C., Frieze, A., Prałat, P.: Some typical properties of the spatial preferred attachment model. Internet Math. **10**, 27–47 (2014)
14. Fortunato, S.: Community detection in graphs. Phys. Rep. **486**(3), 75–174 (2010)
15. Frieze, A., Pérez-Giménez, X., Prałat, P., Reiniger, B.: Perfect matchings and Hamiltonian cycles in the preferential attachment model (preprint)

16. Clauset, A., Newman, M.E.J., Moore, C.: Finding community structure in very large networks. Phys. Rev. E **70**, 066111 (2004)
17. da Costa, L.F., Rodrigues, F.A., Travieso, G., Boas, P.R.U.: Characterization of complex networks: a survey of measurements. Adv. Phys. **56**, 167–242 (2007)
18. Girvan, M., Newman, M.E.: Community structure in social and biological networks. Proc. Natl. Acad. Sci. **99**(12), 7821–7826 (2002)
19. Janssen, J., Hurshman, M., Kalyaniwalla, N.: Model selection for social networks using graphlets. Internet Math. **8**(4), 338–363 (2013)
20. Janssen, J., Prałat, P., Wilson, R.: Geometric graph properties of the spatial preferred attachment model. Adv. Appl. Math. **50**, 243–267 (2013)
21. Janssen, J., Prałat, P., Wilson, R.: Non-uniform distribution of nodes in the spatial preferential attachment model. Internet Math. **12**(1–2), 121–144 (2016)
22. Lancichinetti, A., Fortunato, S.: Limits of modularity maximization in community detection. Phys. Rev. E **84**, 066122 (2011)
23. McDiarmid, C., Skerman, F.: Modularity in random regular graphs and lattices. Electron. Notes Discrete Math. **43**, 431–437 (2013)
24. McDiarmid, C., Skerman, F.: Modularity of tree-like and random regular graphs (preprint)
25. Mihail, M., Papadimitriou, C., Saberi, A.: On certain connectivity properties of the internet topology. In: Proceedings of Conference on Foundations of Computer Science, pp. 28–35 (2003)
26. Montgolfier, F., Soto, M., Viennot, L.: Asymptotic modularity of some graph classes. In: Asano, T., Nakano, S., Okamoto, Y., Watanabe, O. (eds.) ISAAC 2011. LNCS, vol. 7074, pp. 435–444. Springer, Heidelberg (2011). doi:10.1007/978-3-642-25591-5_45
27. Newman, M.E.J.: Assortative mixing in networks. Phys. Rev. Lett. **89**, 208701 (2002)
28. Newman, M.E.J.: Fast algorithm for detecting community structure in networks. Phys. Rev. E **69**, 066133 (2004)
29. Newman, M.E.J.: Power laws, Pareto distributions and Zipf's law. Contemp. Phys. **46**(5), 323–351 (2005)
30. Newman, M.E.J.: The structure and function of complex networks. SIAM Rev. **45**(2), 167–256 (2003)
31. Newman, M.E.J., Girvan, M.: Finding and evaluating community structure in networks. Phys. Rev. E **69**, 026–113 (2004)
32. Pittel, B.: Note on the heights of random recursive trees and random m-ary search trees. Random Struct. Algorithms **5**, 337–347 (1994)
33. Watts, D.J., Strogatz, S.H.: Collective dynamics of 'small-world' networks. Nature **393**, 440–442 (1998)

On Mixing in Pairwise Markov Random Fields with Application to Social Networks

Konstantin Avrachenkov[1]([⊠]), Lenar Iskhakov[2], and Maksim Mironov[2]

[1] Inria Sophia Antipolis, 2004 Route des Lucioles, Sophia-Antipolis, France
k.avrachenkov@inria.fr
[2] Moscow Institute of Physics and Technology, Dolgoprudny, Russia
lenar-iskhakov@yandex.ru, maxim-m94@mail.ru

Abstract. We consider pairwise Markov random fields which have a number of important applications in statistical physics, image processing and machine learning such as Ising model and labeling problem to name a couple. Our own motivation comes from the need to produce synthetic models for social networks with attributes. First, we give conditions for rapid mixing of the associated Glauber dynamics and consider interesting particular cases. Then, for pairwise Markov random fields with submodular energy functions we construct monotone perfect simulation.

1 Introduction

Pairwise Markov random fields or Markov random fields with nonzero potential functions only for cliques of size two have a large number of applications in statistical physics, image processing and machine learning. Let us mention just a few very important particular cases and applications. Ising [9], Potts [13] and Solid-on-Solid (SOS) [12,16] models are the basic models in statistical physics. Metric Markov random fields and the generalized Potts model are very successfully applied in image processing [5,6,18]. Pairwise Markov random fields are also extensively used in the study of classification and labeling problems, see e.g. [4,8,10].

Our own motivation to study pairwise Markov random fields comes from the need to model the distribution of attributes in social networks such as age, gender, interests. The fact that friends or acquaintances in social networks share common characteristics is widely observed in real networks and is referred to as homophily. The property of homophily implies that we expect that the more clustered social network members are, the more likely they are to share same attribute. Nowadays social networks are intensively researched by both sociologists and computer scientists. However, if one wants to check some hypotheses about social networks or to test some algorithm such as a sampling method, one needs a lot of social network examples to consider and to test. In [3] a model of synthetic social network with attributes has been proposed to test subsampling chain-referral methods on many network instances with various properties. The synthetic network model of [3] is similar in spirit to the SOS model and

© Springer International Publishing AG 2016
A. Bonato et al. (Eds.): WAW 2016, LNCS 10088, pp. 127–139, 2016.
DOI: 10.1007/978-3-319-49787-7_11

well represents the distribution of ordinal attributes such as age. Here we study much more general model which could be used to model ordinal as well as non-ordinal attributes' distribution in social networks. Of course, we hope that the results will also be of interest to researchers from statistical physics and machine learning communities.

Specifically, in the present work we consider a general pairwise Markov random field and provide conditions for rapid mixing of the associated Glauber dynamics. Rapid mixing guarantees that we can quickly generate many configurations of attributes corresponding to a given Gibbs distribution or energy function. In the important particular case of submodular energy functions, we go a step further and construct a perfect simulation which samples quickly without bias from the target distribution. Our results significantly generalize the corresponding results for the Ising model, see e.g. [11]. The proof in [11] relies on the particular size and values of the interaction matrix.

Finally, we would like to note that even though our model has some common features with the exponential random graph model (see e.g., [15]), there are important differences between these two models. The exponential random graph model generates the graph, whereas our model assumes that the graph is given and generates a configuration of attributes over the graph.

2   Model

Let a graph $G = (V, E)$, $|V| = n$, be given. In addition, each vertex v has an attribute which takes a value from the finite set $M = \{1, ..., m\}$. We denote by $\sigma \in \Omega = M^n$ a configuration, where each vertex $v \in V$ takes its own certain value $\sigma(v) \in M$ of the attribute. In the present work we restrict ourselves to the model with one attribute. Now we introduce symmetric *interaction* matrix $\mathbb{V}$ of size $m \times m$, and say, that the energy of configuration σ is given by

$$\varepsilon(\sigma) = \sum_{\{v_1, v_2\} \in E} \mathbb{V}(\sigma(v_1), \sigma(v_2)).$$

Let us call $|\mathbb{V}|$ the maximum absolute value of matrix $\mathbb{V}$ elements. Next we consider *Gibbs distribution* with respect to the introduced energy:

$$\pi^*(\sigma) = \frac{e^{-\beta \varepsilon(\sigma)}}{\sum_{\tau \in M^G} e^{-\beta \varepsilon(\tau)}} = Z^{-1}(\beta) e^{-\beta \varepsilon(\sigma)},$$

where $\beta = \frac{1}{T}$ is some parameter, the inverse temperature of the system, and $Z(\beta)$ is the normalizing constant or, in statistical physics terminology, the partition function. This distribution describes the *pairwise Markov random field* over graph G. We shall also refer to this distribution as *network attribute distribution*.

We would like to sample configurations from the distribution $\pi^*(\sigma)$ to test various algorithms on a series of network realisations. However, the main problem is that the probability space is enormous and it is impossible to sample from

Gibbs distribution without additional techniques. One such technique is Glauber dynamics, described just below and another technique is monotone perfect simulations described in detail in Sect. 5.

Let $\mathcal{N}(v)$ be the set of neighbours of vertex v. Then, we define the *local energy* $\varepsilon_i(\sigma, v)$ for vertex v with respect to value i in configuration σ as follows:

$$\varepsilon_i(\sigma, v) = \sum_{u \in \mathcal{N}(v)} \mathbb{V}(i, \sigma(u)).$$

This formula calculates energy in the neighbourhood of v provided that the value of the attribute for v was updated to i. Then, we call the *local distribution* for vertex v in configuration σ the probability distribution on set $\{1, 2, ..., m\}$ with respect to the local energy:

$$p_i(\sigma, v) = \mathbb{P}(\sigma(v) \to i) := \frac{e^{-\beta \varepsilon_i(\sigma, v)}}{\sum_{k \in M} e^{-\beta \varepsilon_k(\sigma, v)}} = Z^{-1}(\sigma, v, \beta) \cdot e^{-\beta \varepsilon_i(\sigma, v)},$$

which is the probability to update value in v to i.
The *Glauber dynamics* is defined as follows:

1. Choose arbitrary starting distribution π^0 and then choose values for vertices according to π^0;
2. Choose uniformly random vertex v;
3. Update value for v according to the local distribution;
4. Go to step 2.

Let us denote by $\mathcal{X} = \{X_t, t \geq 0\}$ the Markov chain associated with the Glauber dynamics, with starting distribution π^0 and transition matrix $P = \{P_{\sigma,\tau}\}_{\sigma,\tau \in \Omega}$, $P_{\sigma,\tau} = \mathbb{P}\{X_{t+1} = \tau | X_t = \sigma\}$, which is associated with steps 2–3. If steps 2–3 are repeated t times, π^t will stand for the distribution on space of configurations at time moment t. Sometimes we.shall also use $P_\sigma^t(\cdot)$ to denote the probability distribution of $\mathcal{X}$ on Ω at time moment t to emphasize that $\mathcal{X}$ starts from certain configuration σ.

Before we proceed further, let us notice that the introduced model implies some well-known particular cases. For example,

$$\mathbb{V} = \begin{pmatrix} 1 & 0 & \cdots & 0 \\ 0 & 1 & \cdots & 0 \\ \vdots & \vdots & \ddots & \vdots \\ 0 & 0 & \cdots & 1 \end{pmatrix}$$

corresponds to the Potts model. If $m = 2$, then the Potts model becomes the Ising model. If now we take $\mathbb{V}(i, j) = f(|i - j|)$ with some convex function $f(\cdot)$, we obtain the metric Markov random field model extensively used in image processing. In [3], the Markov random field with quadratic $f(\cdot)$ was used to model social networks with ordinal attributes. The case $\mathbb{V}(i, j) = |i - j|$ corresponds to the SOS model.

3   Preliminaries

Here we give several background results, which we will use in sequel. It is well-known, see e.g., [7,11], that the Markov chain $\mathcal{X}$ corresponding to the Glauber dynamics is reversible with the stationary distribution π^*.

Lemma 1. *Markov chain $\mathcal{X}$ is time-reversible with the stationary distribution given by $\pi^*(\sigma) = Z^{-1}(\beta)e^{-\beta\varepsilon(\sigma)}$. In other words,*

$$\pi^*(\sigma) \cdot P_{\sigma,\tau} = \pi^*(\tau) \cdot P_{\tau,\sigma},$$

for all $\sigma, \tau \in \Omega$.

For two distributions π_1, π_2 on state space Ω we define the *total variation* distance between them as

$$||\pi_1 - \pi_2||_{TV} = \frac{1}{2} \sum_{\sigma \in \Omega} |\pi_1(\sigma) - \pi_2(\sigma)|.$$

Let μ and ν be two distributions on the same state space Ω. Pair of random variables (X_μ, X_ν) forms *coupling*, if it is distributed such that marginal distribution of X_μ is μ and marginal distribution of X_ν is ν. The main motivation for introducing such term is the following lemma [7].

Lemma 2. *Let ν and μ be two probability distributions on Ω. Then*

$$||\mu - \nu||_{TV} = \inf\{\mathbb{P}(X_\mu \neq X_\nu) \mid (X_\mu, X_\nu) \text{ is a coupling of } \mu \text{ and } \nu\}.$$

This lemma is very useful, because a comparison between distributions is reduced to comparison between random variables.

Here is one more lemma, which shows how the total variation distance from the stationary distribution can be estimated [7,11].

Lemma 3. *Let σ and τ be initial configurations from state space Ω. Then*

$$||\pi^t - \pi^*||_{TV} \leqslant \max_{\sigma,\tau \in \Omega} ||P_\sigma^t(\cdot) - P_\tau^t(\cdot)||_{TV}.$$

Now we introduce metric on configuration space Ω. Let $\rho(\cdot, \cdot)$ by definition be equal to

$$\rho(\sigma, \tau) = \sum_{v \in V} |\sigma(v) - \tau(v)|.$$

Lemma 4. *Let α be such that for every two neighbor configurations σ, $\tau(\rho(\sigma, \tau) = 1)$ corresponding random values X_σ^1 and X_τ^1 satisfy an inequality*

$$\mathsf{E}\rho(X_\sigma^1, X_\tau^1) \leqslant e^{-\alpha}.$$

Then

$$\forall t \in \mathbb{N}, \ \forall \sigma, \tau \in \Omega \rightarrow \mathsf{E}(\rho(X_\sigma^t, X_\tau^t)) \leqslant \mathsf{diam}(\Omega) \cdot e^{-\alpha t}.$$

Lemma 4 shows how the introduced property can be generalized from neighbor configurations to the whole space Ω for an arbitrary time moment.

For some $\varepsilon > 0$, the *mixing time* is defined as follows:

$$t_{mix}(\varepsilon) = \min(t \in \mathbb{N} \mid ||\pi^t - \pi||_{TV} < \varepsilon).$$

Next lemma is based on Lemma 4 and it provides an upped bound for the mixing time with respect to α.

Lemma 5. *Suppose $\alpha > 0$ is such that $\mathsf{E}(\rho(X_\sigma^1, X_\tau^1)) \leqslant e^{-\alpha}$ for all neighbour configurations σ, τ. Then*

$$t_{mix} \leqslant \left\lceil \frac{1}{\alpha}[\ln(\mathsf{diam}(\Omega)) + \ln(1/\varepsilon)] \right\rceil.$$

Lemmas 4 and 5 are borrowed from [11]. Actually, for our following results it would be enough to refer only to Lemma 5. But we mention here intermediate steps to help a reader to better understand the proof of our main result.

4   Main Results

We can now formulate the main result of this article which says that under certain conditions the Glauber dynamics corresponding to the general pairwise Markov random fields mixes rapidly.

Theorem 1. *Let $\triangle$ be the maximum degree of graph $G = (V, E)$, $|V| = n$ and $\mathbb{V}$ be the interaction matrix. Let also β be the inverse temperature and $M = \{1, 2, ..., m\}$ be the set of attribute values. If*

$$\beta < \frac{1}{4|\mathbb{V}|} \ln\left(1 + \frac{1}{\triangle m}\right),$$

then

$$t_{mix} \leqslant \left\lceil \frac{n(\ln(n) + \ln(m-1) + \ln(\frac{1}{\varepsilon}))}{1 - \triangle m(e^{4\beta|\mathbb{V}|} - 1)} \right\rceil.$$

We would like to notice that independently from temperature the mixing time is at least of order $n \ln(n)$. It is so, because achieving stationary distribution by iterating means that every vertex of the graph has to be updated at least once. As n grows to infinity, we must do order $n \ln(n)$ Markov chain steps to make the probability of updating each vertex at least once tending to 1. More details on various lower bounds can be found in [11].

Before we proceed to prove the theorem, let us also notice that it claims that the upper bound is of order $n \log n$. The corresponding result for the Ising model has been shown in e.g., [11]. The present extension is not straightforward, since the proof in [11] is based on the particular form of the interaction matrix $\mathbb{V}$.

Proof. Let us choose two arbitrary configurations σ and τ at time 0 and say that random vectors X_σ^t and X_τ^t have distributions $P_\sigma^t(\cdot)$ and $P_\tau^t(\cdot)$, respectively. Then define $pref_k(\sigma, w), k \leqslant m$, as the *prefix sum* of probabilities to label w with one of the first k attribute values at the next step, namely,

$$pref_k(\sigma, w) = \sum_{i=1}^{k} p_i(\sigma, w).$$

Let us consider the following probability distribution of pair (X_σ^t, X_τ^t): first we uniformly at random choose a vertex w to update (common for both configurations) and then we choose uniformly at random a value U from $[0, 1]$. Then we set new configurations $X_\phi^t(U, w), \phi \in \{\sigma, \tau\}$ at time t by the relation

$$X_\phi^t(U, w)(\overline{w}) = \begin{cases} \phi(\overline{w}) & \overline{w} \neq w \\ \min(k|pref_k(\phi, w) \geqslant U) & \overline{w} = w \end{cases}. \tag{1}$$

where function $X_\phi^1 : [0, 1] \times V \to \Omega$ becomes a random vector, if U and w are random variables.

It is easy to see that distribution of pair $(X_\sigma^t(U, w), X_\tau^t(U, w))$ is coupling for $P_\sigma^t(\cdot)$ and $P_\tau^t(\cdot)$.

Then, we are going to find an $\alpha > 0$ from Lemma 4 for two neighbor configurations. Let σ, τ be two neighbor configurations with unique difference in vertex v, i.e., $|\sigma(v) - \tau(v)| = 1$. Let also w be a uniformly chosen random vertex. If $w = v$, then

$$\rho(X_\sigma^1(U, w), X_\tau^1(U, w)) = 0.$$

If $w \notin \mathcal{N}(v) \cup \{v\}$, then

$$\rho(X_\sigma^1(U, w), X_\tau^1(U, w)) = |\sigma(v) - \tau(v)| = 1.$$

It is so, because in both cases local distributions for w are the same for both configurations. And if $w \in \mathcal{N}(v)$, then

$$\rho(X_\sigma^1(U, w), X_\tau^1(U, w)) = |\sigma(v) - \tau(v)| + |X_\sigma^1(U, w)(w) - X_\tau^1(U, w)(w)|.$$

According to probabilities of each case, we can write

$$\mathsf{E}\rho(X_\sigma^1(U, w), X_\tau^1(U, w)) = 1 - \frac{1}{n} + \frac{1}{n} \cdot \sum_{w \in \mathcal{N}(v)} \mathsf{E}|X_\sigma^1(U, w)(w) - X_\tau^1(U, w)(w)|. \tag{2}$$

Thus, an upper bound for the sum in (2) is needed. The following lemma helps to achieve the result and is the key element of this work.

Lemma 6. *For arbitrary $\sigma, \tau \in \Omega$ and for all $w \in V$ the following equation holds*

$$\mathsf{E}|X_\sigma^1(U, w)(w) - X_\tau^1(U, w)(w)| = \sum_{i=1}^{m} |pref_i(\sigma, w) - pref_i(\tau, w)|. \tag{3}$$

Proof. The expectation in (3) is based on uniform random variable U distributed on $[0, 1]$. Let us place on segment $[0, 1]$ precisely m red points that correspond to $pref_i(\sigma, w)$ and m blue points that correspond to $pref_i(\tau, w)$, $1 \leqslant i \leqslant m$. Since $pref_m(\sigma, w) = pref_m(\tau, w) = 1$, we have $2m - 1$ disjoint (with no common internal points) subsegments with red or blue endpoints (some subsegments may have length 0), they form a set $\{l_k\}_{k=1}^{2m-1}$. Let subsegment l_k have a value $h_{\sigma,k}$, if $h_{\sigma,k}$ satisfies $l_k \subset [pref_{h_{\sigma,k}-1}(\sigma, v), pref_{h_{\sigma,k}}(\sigma, w)]$. Thus, by definition the mean of $|X_\sigma^1(U, w)(w) - X_\tau^1(U, w)(w)|$ is

$$\mathsf{E}|X_\sigma^1(U, w)(w) - X_\tau^1(U, w)(w)| = \sum_{k=1}^{2m-1} \text{length}(l_k) \cdot |h_{\sigma,k} - h_{\tau,k}|.$$

In other words, the length of l_k appears in the expectation as many times as the difference between the values of the attribute for updates in σ and τ. Therefore, we now calculate the number of times that the length of each subsegment is added to the result in the right hand side of the above equality. Towards this goal, for the moment let us fix k and let $h_{\sigma,k} = a$, $h_{\tau,k} = b$ and without loss of generality $b \geqslant a$. Thus, the following series of inequalities hold

$$\begin{cases} pref_a(\sigma, w) \geqslant pref_a(\tau, w), \\ pref_{a+1}(\sigma, w) \geqslant pref_{a+1}(\tau, w), \\ \dots \\ pref_b(\sigma, w) \geqslant pref_b(\tau, w). \end{cases}$$

Let us identify terms $|pref_i(\sigma, w) - pref_i(\tau, w)|$ in (3) which contain the contribution from the subsegment l_k. The length of l_k is added for the first time in the right hand side of (3) for $i = a$, because according to the definition of a the minimum i such that segment $[0, pref_i(\sigma, w)]$ contains l_k is $i = a$, meantime $pref_a(\tau, w)$ does not contain this subsegment. Second time it is added for $i = a + 1$ and so on, the last time it is added for $i = b - 1$, which comes from definition of b. Hence, l_k is added exactly $b - a$ times. This establishes equivalence between the sums and completes the proof of the lemma. $\qquad \square$

Actually, this lemma will be used only for neighbor configurations σ, τ, as it was mentioned before Lemma 6. Recall that Lemma 4 and then Lemma 5 give us an upper bound on the mixing time, but to apply them we need to obtain the corresponding inequalities on neighbour configurations. Therefore, we give a uniform upper bound for (3). For convenience we introduce

$$S_i = \sum_{u \in \mathcal{N}(w) \setminus \{v\}} \mathbb{V}(i, \sigma(u)) = \sum_{u \in \mathcal{N}(w) \setminus \{v\}} \mathbb{V}(i, \tau(u)),$$

$$a_i = \exp\left(-\beta \sum_{u \in \mathcal{N}(w)} \mathbb{V}(i, \sigma(u))\right) = \exp\left(-\beta(S_i + \mathbb{V}(i, \sigma(v)))\right),$$

$$b_i = \exp\left(-\beta \sum_{u \in \mathcal{N}(w)} \mathbb{V}(i, \tau(u))\right) = \exp\left(-\beta(S_i + \mathbb{V}(i, \tau(v)))\right).$$

Thus,

$$\begin{cases} p_i(\sigma, w) = \frac{a_i}{a_1 + \ldots a_m} \\ p_i(\tau, w) = \frac{b_i}{b_1 + \ldots + b_m} \end{cases}.$$

The following inequality will be useful:

$$\frac{a_i b_k}{a_k b_i} = \exp(-\beta(\mathbb{V}(i, \sigma(v)) + \mathbb{V}(k, \tau(v)) - \mathbb{V}(k, \sigma(v)) - \mathbb{V}(i, \tau(v)))) \leqslant e^{4\beta|\mathbb{V}|}. \quad (4)$$

Then, the upper bound for (3) can be derived as follows:

$$\sum_{k=1}^{m} |pref_k(\sigma, w) - pref_k(\tau, w)| \leqslant \sum_{k=1}^{m} \sum_{i=1}^{k} |p_i(\sigma, w) - p_i(\tau, w)|$$

$$\leqslant m \sum_{i=1}^{m} |p_i(\sigma, w) - p_i(\tau, w)| = m \sum_{i=1}^{m} \left| \frac{a_i}{a_1 + \ldots + a_m} - \frac{b_i}{b_1 + \ldots + b_m} \right|$$

$$\leqslant \frac{m}{(a_1 + \ldots + a_m)(b_1 + \ldots + b_m)} \sum_{i=1}^{m} |a_i(b_1 + \ldots + b_m) - b_i(a_1 + \ldots + a_m)|$$

$$\leqslant \frac{m}{(a_1 + \ldots + a_m)(b_1 + \ldots + b_m)} \sum_{i=1}^{m} \sum_{j=1}^{m} |a_i b_j - a_j b_i|$$

$$\leqslant \frac{m}{(a_1 + \ldots + a_m)(b_1 + \ldots + b_m)} \sum_{i=1}^{m} \sum_{j=1}^{m} a_j b_i \left| e^{4\beta|\mathbb{V}|} - 1 \right| \leqslant m \left(e^{4\beta|\mathbb{V}|} - 1 \right). \quad (5)$$

And now collecting together (2), (3) and (5), we obtain

$$\mathsf{E}\rho(X_\sigma^1, X_\tau^1) \leqslant 1 - \frac{1 - \triangle m e^{4\beta|\mathbb{V}|}}{n} \leqslant \exp\left(-\frac{1 - \triangle m (e^{4\beta|\mathbb{V}|} - 1)}{n}\right). \quad (6)$$

Indeed, the diameter of Ω is equal to $n(m-1)$ and it corresponds to the distance between configurations $\hat{1} = (1, 1, ..., 1)$ and $\hat{m} = (m, m, ..., m)$. Now invoking Lemma 5 with α provided by (6), we obtain the upper bound for $t_{mix}(\varepsilon)$ given in the theorem statement. □

Once we proved the theorem, we can think about modifications of the interaction matrix $\mathbb{V}$ and their influence on the model. It is easy to see from the definition of the Gibbs distribution that if we consider matrix $c\mathbb{V}$, where each element of matrix $\mathbb{V}$ is multiplied by a factor c, we obtain a new probability distribution on the configuration space Ω which is actually equal to the Gibbs distribution for the pair $\mathbb{V}$ and $c \cdot \beta$. Moreover, if we add some constant d to all elements of matrix $\mathbb{V}$, then the distribution will not change at all. Now we notice that $|\mathbb{V}|$ is mentioned in Theorem 1 and we can diminish it to some extent. This results in the following refinement.

Corollary 1. *Let Δ be the maximum degree of graph $G = (V, E)$, $|V| = n$ and $\mathbb{V}$ be the interaction matrix. Let also β be the inverse temperature and $M = \{1, 2, ..., m\}$ be the set of attribute values. Let also*

$$K = \frac{\max_{x,y} \mathbb{V}(x, y) - \min_{x,y} \mathbb{V}(x, y)}{2}.$$

If

$$\beta < \frac{1}{4K} \ln\left(1 + \frac{1}{\Delta m}\right),$$

then

$$t_{mix} \leq \left\lceil \frac{n(\ln(n) + \ln(m-1) + \ln(\frac{1}{\varepsilon}))}{1 - \Delta m(e^{4\beta K} - 1)} \right\rceil.$$

This refinement gives a slightly better bound for the mixing time. However, we prefer to keep both formulations since the first variant could be just more notationally convenient in some setting.

In the case of quadratic dependencies in $\mathbb{V}$ we obtain even better upper bound.

Theorem 2. *If $\mathbb{V}(x, y) = (x - y)^2$, and*

$$\beta < \frac{1}{2(m-1)} \ln\left(1 + \frac{1}{\Delta m}\right),$$

then

$$t_{mix} \leq \left\lceil \frac{n(\ln(n) + \ln(m-1) + \ln(\frac{1}{\varepsilon}))}{1 - \Delta m(e^{2\beta(m-1)} - 1)} \right\rceil.$$

In this particular case $|\mathbb{V}| = (m-1)^2$ and the above mentioned result is obviously more efficient than the one which can be obtained from Corollary 1.

Proof. The only difference in the proof of this theorem with respect to the previous results is in inequality (4). Recall that we use that inequality only for neighbour configurations σ and τ, which means that there is a vertex v such that σ and τ agree everywhere but in vertex v, and for that vertex it holds that $|\sigma(v) - \tau(v)| = 1$. Since $\mathbb{V}(x, y) = (x - y)^2$, we can rewrite the right hand side of inequality (4) in the following way:

$$\frac{a_i b_k}{a_k b_i} = \exp(-\beta((i - \sigma(v))^2 + (k - \tau(v))^2 - (k - \sigma(v))^2 - (k - \tau(v))^2)),$$

Now, without loss of generality $\sigma(v) + 1 = \tau(v)$, and then

$$\frac{a_i b_k}{a_k b_i} = \exp(2\beta(k - i)) \leq \exp(2\beta(m - 1)). \tag{7}$$

The latter provides us α for Lemma 5 and leads to the proof of the theorem. $\square$

Remark. All three results mentioned above show that there is fast mixing with respect to some condition on the temperature of the system. Actually, it is impossible to proof fast mixing in general case independently of the temperature. It is already shown for the Ising model, and we can generalize that fact and can demonstrate that for arbitrary m and $m \times m$ matrix $\mathbb{V}$, where not all elements are equal, there exists a temperature and a graph such that mixing time has exponential order in terms of graph size. Moreover, we believe, that for every m and V there exists an example of a graph such that mixing is fast independently of the temperature. This is a good question to address in future research.

5   Simulations

5.1   Monotone Perfect Markov Chain Monte Carlo

In this section we are about to compare theoretical result with real simulations. Of course, for simulation one can just run the Glauber dynamics and use the bounds on the mixing time from Theorem 1 or Corollary 1 to indicate the simulation stopping time. However, if matrix $\mathbb{V}$ has some structure, it appears to be possible to construct a monotone perfect Markov Chain Monte Carlo (MCMC) simulation which produces perfect sampling and has a natural stopping rule. Our construction is based on the general recommendations given in [14]. Towards this end, under coupling described by Eq. (1), we need to show that for any two configurations σ and τ, such that $\sigma \preceq \tau$, we have $X_\sigma^t(U, w) \preceq X_\tau^t(U, w)$, where the order $\preceq$ means that for all vertices $v \in V$ it holds that $\sigma(v) \leqslant \tau(v)$. Unfortunately, this is true not for any matrix $\mathbb{V}$ and here, unlike in Theorem 1, we have to impose additional restrictions on $\mathbb{V}$.

Let us call matrix $\mathbb{V}$ *submodular* if for all $i < j, k < l$ it holds that

$$\mathbb{V}(i, k) + \mathbb{V}(j, l) \leqslant \mathbb{V}(i, l) + \mathbb{V}(j, k).$$

For example, matrix $\mathbb{V}(x, y) = f(x - y)$ is *submodular*, when f is a convex function (in particular, the matrix $\mathbb{V}$ in Theorem 2 is submodular).

Lemma 7. *Let $\sigma \preceq \tau$ and there is a coupling defined by equality (1) for submodular matrix $\mathbb{V}$. Then*

$$X_\sigma^t(U, w) \preceq X_\tau^t(U, w).$$

Proof. Suppose $t = 1$. Since the introduced order is transitive, we can limit consideration to neighbor configurations. So, let $\sigma(u) = \tau(u)$ for all $u \in V \setminus \{v\}$ and $\sigma(v) + 1 = \tau(v)$. Let some vertex w be chosen for update. If $w \notin \mathcal{N}(v)$ then the neighborhood of w is the same for both configurations and it holds that $X_\sigma^1(U, w)(w) = X_\tau^1(U, w)(w)$. Then, consider $w \in N(v)$. It will be enough to prove that for all $k \leqslant m$ the following inequality holds

$$pref_k(\sigma, w) \leqslant pref_k(\tau, w)$$

to be sure that

$$X_\sigma^1(U, w)(w) = \min(k | pref_k(\sigma, w) \geqslant U) \leqslant \min(k | pref_k(\tau, w) \geqslant U) = X_\tau^1(U, w)(w).$$

Here we will use notations of Lemma 6.

$$pref_k(\sigma, w) - pref_k(\tau, w) = \sum_{i=0}^{k} p_i(\sigma, w) - \sum_{i=0}^{k} p_i(\tau, w)$$

$$= \sum_{i=0}^{k} \frac{a_i}{a_0 + \ldots + a_m} - \sum_{i=0}^{k} \frac{b_i}{b_0 + \ldots + b_m}$$

$$= \frac{(a_0 + \ldots + a_k) \cdot (b_0 + \ldots + b_m) - (a_0 + \ldots + a_m) \cdot (b_0 + \ldots + b_k)}{(a_0 + \ldots + a_m)(b_0 + \ldots + b_m)}$$

$$= \frac{(a_0 + \ldots + a_k) \cdot (b_{k+1} + \ldots + b_m) - (a_{k+1} + \ldots + a_m) \cdot (b_0 + \ldots + b_k)}{(a_0 + \ldots + a_m)(b_0 + \ldots + b_m)}$$

$$= \frac{1}{(a_0 + \ldots + a_m)(b_0 + \ldots + b_m)} \sum_{i \leqslant k < j}^{m} (a_i b_j - a_j b_i) \leqslant 0.$$

The last inequality holds since each summand is at most zero: it is provided by Eq. (4), submodular property of matrix $\mathbb{V}$ and the fact that summation is performed with $i < j$. By induction argument the proof immediately extends for arbitrary t.   □

Now we can propose the following algorithm:

Algorithm 1. Monotone perfect MCMC

$U_t \leftarrow$ random uniform variables from the segment [0,1]
$w_t \leftarrow$ random uniform variables from the set V
$T \leftarrow 1$
repeat
    $upper \leftarrow \hat{m}$
    $lower \leftarrow \hat{1}$
    for $t = -T \ldots -1$ **do**
        $upper \leftarrow X^1_{upper}(U_t, w_t)$
        $lower \leftarrow X^1_{lower}(U_t, w_t)$
    $T \leftarrow 2T$
until $upper = lower$
return $upper, T$

It is needed to say that the algorithm uses the same random pair (U_t, w_t) at the same t, that is why we initialize them only once during the first call. The required number of steps for this algorithm is upper bounded by $4T_*$, where T_* is the smallest T such that $upper$ and $lower$ values converge. In this case T_* is a random value depending on U_t and w_t. Having found T such that $T < T_* \leqslant 2T$ one can make a binary search to find out the accurate value of T_*. This calculation has asymptotic complexity of order $T_* \ln T_*$.

According to [14], we have:

$$\mathsf{E}T_* \leqslant 2t_{mix} \cdot (1 + \ln n + \ln m).$$

This gives an idea that the Glauber dynamics and Monotone perfect MCMC are comparable in terms of computational requirements. Of course, the advantage of the monotone perfect MCMC is that it produces sampling from the exact stationary distribution.

5.2 Numerical Example with Real Network

Let us consider well-known social network with attributes *AddHealth* [1]. For our experiments, we take as attribute the grade (class) of a pupil at school. It is an ordinal attribute in the interval between 7 and 12. It seems natural that this network has cluster structure based on class attribute, because the probability of friendship between two pupils is bigger if their classes are not so far apart in time. For this purpose, as in [3], we have chosen 6×6 interaction matrix $\mathbb{V}(x, y) = (x - y)^2$. Since $\mathbb{V}$ is submodular, we can use monotone perfect MCMC. We have taken publically available AddHealth graph [2] with the number of vertices $n = 1996$ and with the maximum degree $\triangle = 36$. In this case Theorem 2 provides fast mixing for $\beta < 0.000461895$, or equivalently, for the temperature >2165.

If we choose $\beta = 0.0002$, Theorem 2 gives the upper bound 27000 on the mixing time while perfect MCMC algorithm makes about 20000–25000 running steps. Moreover, if we choose β bigger than provided by Theorem 2, e.g., about 0.04, the perfect MCMC is still fast enough finishing approximately after 200000 steps. Since we have a relation between the expectation of the number of steps in perfect MCMC and the mixing time, we realize that, on the one hand, our theorem is in agreement with experiment and, on the other hand, on that particular graph there is fast mixing on broader set of parameters. The question if it is possible to obtain a tighter mixing time estimate is an interesting direction for future research.

We have also tried to fit the value of β for the AddHealth data using a variation of the method of moments (see e.g., [17]). Specifically, we tried to fit the simulated energy to the energy of the AddHealth data, which is equal to 12328. The perfect simulation algorithm converges in acceptable time for β as high as 0.125, which gives the energy level around 15000. We think it is a reasonable match. It is interesting that AddHealth social network is on the boundary of rapid mixing. This might not be a coincidence as a social network can self-organize to find a balance between sufficiently rapid mixing and division into communities.

References

1. The National Longitudinal Study of Adolescent to Adult Health. http://www.cpc.unc.edu/projects/addhealth
2. Freeman, L.C.: Social Networks Datasets. University of California, Irvine. http://moreno.ss.uci.edu/data.html

3. Avrachenkov, K., Neglia, G., Tuholukova, A.: Subsampling for chain-referral methods. In: Wittevrongel, S., Phung-Duc, T. (eds.) ASMTA 2016. LNCS, vol. 9845, pp. 17–31. Springer, Heidelberg (2016). doi:10.1007/978-3-319-43904-4_2
4. Basu, S., Bilenko, M., Mooney, R.J.: A probabilistic framework for semi-supervised clustering. In: Proceedings of the 10th ACM SIGKDD, pp. 59–68 (2004)
5. Boykov, Y., Veksler, O., Zabih, R.: Markov random fields with efficient approximations. In: Proceedings of Computer Vision and Pattern Recognition, pp. 648–655 (1998)
6. Boykov, Y., Veksler, O., Zabih, R.: Fast approximate energy minimization via graph cuts. IEEE Trans. Pattern Anal. Mach. Intell. **23**(11), 1222–1239 (2001)
7. Brémaud, P.: Markov Chains: Gibbs Fields, Monte Carlo Simulation, and Queues, Texts in Applied Mathematics, vol. 31. Springer, New York (1998)
8. Chakrabarti, S., Dom, B., Indyk, P.: Enhanced hypertext categorization using hyperlinks. ACM SIGMOD Rec. **27**(2), 307–318 (1998)
9. Ising, E.: Beitrag zur theorie des ferromagnetismus. Z. Phys. A: Hadrons Nucl. **31**(1), 253–258 (1925)
10. Kleinberg, J., Tardos, E.: Approximation algorithms for classification problems with pairwise relationships: metric labeling and Markov random fields. J. ACM **49**(5), 616–639 (2002)
11. Levin, D.A., Peres, Y., Wilmer, E.L.: Markov Chains and Mixing Times. American Mathematical Society, Providence (2009)
12. Mazel, A.E., Suhov, Y.M.: Random surfaces with two-sided constraints: an application of the theory of dominant ground states. J. Stat. Phys. **64**(1–2), 111–134 (1991)
13. Potts, R.B.: Some generalized order-disorder transformations. Math. Proc. **48**(1), 106–109 (1952)
14. Propp, J.G., Wilson, D.B.: Exact sampling with coupled Markov chains and applications to statistical mechanics. Random Struct. Algorithms **9**(1–2), 223–252 (1996)
15. Robins, G., Pattison, P., Kalish, Y., Lusher, D.: An introduction to exponential random graph (p*) models for social networks. Soc. Netw. **29**(2), 173–191 (2007)
16. Rozikov, U.A., Suhov, Y.M.: Gibbs measures for SOS models on a Cayley tree. Infin. Dimens. Anal. Quantum. Probab. Relat. Top. **9**(3), 471–488 (2006)
17. Snijders, T.A.B.: The statistical evaluation of social network dynamics. Sociol. Methodol. **31**, 361–395 (2001)
18. Szeliski, R., Zabih, R., Scharstein, D., Veksler, O., Kolmogorov, V., Agarwala, A., Tappen, M., Rother, C.: A comparative study of energy minimization methods for Markov random fields with smoothness-based priors. IEEE Trans. Pattern Anal. Mach. Intell. **30**(6), 1068–1080 (2008)

Subgraphs in Non-uniform Random Hypergraphs

Megan Dewar[1], John Healy[1], Xavier Pérez-Giménez[2(✉)],
Paweł Prałat[2], John Proos[1], Benjamin Reiniger[2], and Kirill Ternovsky[2]

[1] The Tutte Institute for Mathematics and Computing, Ottawa, ON, Canada
[2] Department of Mathematics, Ryerson University, Toronto, ON, Canada
xperez@ryerson.ca

Abstract. Myriad problems can be described in hypergraph terms. However, the theory and tools are not sufficiently developed to allow most problems to be tackled directly within this context. The main purpose of this paper is to increase the awareness of this important gap and to encourage the development of this formal theory, in conjunction with the consideration of concrete applications. As a starting point, we concentrate on the problem of finding (small) subhypergraphs in a (large) hypergraph. Many existing algorithms reduce this problem to the known territory of graph theory by considering the 2-section graph. We argue that this is not the right approach, neither from a theoretical point of view (by considering a generalization of the classic model of binomial random graphs to hypergraphs) nor from a practical one (by performing experiments on two datasets).

Keywords: Random graphs · Random hypergraphs · Subgraphs · Subhypergraphs

1 Introduction

Myriad problems can be described in hypergraph terms. However, the theory and tools are not sufficiently developed to allow most problems to be tackled directly within this context. Hypergraphs are of particular interest in the field of knowledge discovery, where most problems currently modelled as graphs would be more accurately modelled as hypergraphs. Researchers in the knowledge discovery field are particularly interested in the generalization of the concepts of modularity and diffusion to hypergraphs. Such generalizations require a firm theoretical basis on which to develop these concepts. Unfortunately, although hypergraphs were formally defined in the 1960s (and various realizations of hypergraphs were studied long before that), the general formal theory is not as mature as required for the applications of interest to many industry partners or governments. The main purpose of this paper is to increase the awareness of this important gap and to encourage the development of this formal theory, in conjunction with the consideration of concrete applications.

A. Bonato et al. (Eds.): WAW 2016, LNCS 10088, pp. 140–151, 2016.
DOI: 10.1007/978-3-319-49787-7_12

In order to illustrate the issue, let us consider the following "toy example." Consider the coauthorship hypergraph in which vertices correspond to researchers and each hyperedge consists of the set of authors listed on a scientific paper. We have two goals for this dataset. As a first goal, we would like to determine the Erdős number of every researcher (zero for Erdős, one for coauthors of Erdős, two for coauthors of coauthors of Erdős, etc.). Our second goal is to find a minimum set of authors who between them cover all the papers in the subhypergraph consisting only of the seminal papers in a particular field.

Often even though a dataset is naturally represented as a hypergraph we do not work directly on the hypergraph. Instead we reduce the hypergraph to its 2-section graph (the 2-section graph of a hypergraph is obtained by making each hyperedge a clique; see Sect. 2 for a formal definition) or a weighted version of the 2-section. Taking a 2-section of a hypergraph loses some of the information about hyperedges of size greater than 2. Sometimes losing this information does not affect our ability to answer the questions of interest. For example, the Erdős number of an author is the minimum distance between the author's vertex and Erdős' vertex in the hypergraph and this distance is not changed by taking the 2-section. Other times the information lost when taking the 2-section prevents us answering the question of interest. This is the case for our second goal of finding a minimum set of authors that cover a set of papers. In the hypergraph this goal means finding a minimum set of vertices that are incident with every hyperedge of interest. However, taking the 2-section of the hypergraph loses the information about the set of papers that a particular author covers. In fact, the 2-section does not even retain how many papers exist. Basically, if the composition of the hyperedges of size greater than 2 is important in solving a problem, then solving the problem in the 2-section is going to be difficult or impossible.

Besides the information loss, there is another potential downside to working with the 2-section of a hypergraph. Namely, that the 2-section can be much denser than the hypergraph since a single hyperedge of size k implies $\binom{k}{2}$ edges in the 2-section. Depending on the dataset and algorithm being executed the increased density of the 2-section can have a significant detrimental effect on the runtime.

In this paper, we shall be interested in finding subhypergraphs in hypergraphs. While the composition of the hyperedges of size greater than 2 matters when answering this question, it is natural to ask whether 2-section graphs can be used to help answer the question. That is, when determining whether or not a hypergraph H contains H_1 as a subhypergraph, is it useful to look for G_{H_1}, the 2-section of H_1, in G_H, the 2-section of H? Clearly there are many ways that G_{H_1} could appear in G_H without H_1 appearing in H. An obvious technique would be to use the existing graph theoretical tools to find all copies of G_{H_1} in G_H and then simply inspect them, one by one, in the original hypergraph. So perhaps reducing the hypergraph to its 2-section can be used to solve the problem. Maybe in most networks that are considered in practice, any two subhypergraphs inducing the same graph in the 2-section occur with the same probability? This would be desirable, as it would mean that the above technique does not waste a lot of time dealing with subhypergraphs that we are not inter-ested in finding. Of course, even if the 2-section can be used in this way for finding subhypergraphs, the increased density of the 2-section may lead to the graph theoretical tools used being quite inefficient.

In order to deal with the question of the false positive rate of G_{H_1} in G_H, we introduce a natural generalization of Erdős-Rényi (binomial) random graphs to non-uniform random hypergraphs. We study (rigorously, via theorems with proofs) occurrences of a given hypergraph in the random hypergraph. One of the implications of our work is that two hypergraphs H_1, H_2 that induce the same subgraph in the 2-section can have drastically different thresholds for appear-ance. This suggests that the answer to the latest question is "no," and that we have lost something by considering only the 2-section. Assuming that hyperedges in the network we try to analyze occur randomly, our theorems imply that there might be very few (if any) copies of H_1 (the hypergraph we are looking for in the network) but plenty of copies of H_2 (the hypergraph we do not care about). So the algorithm discovers a lot of potential candidates but none of them is what we are looking for!

We investigate two real-world networks: an email hypergraph and the coau-thorship hypergraph that was already mentioned. Not surprisingly, we confirm that hypergraphs that are not distinguishable in the 2-section graph occur with different probabilities (as predicted by the model). Hence we feel that using existing graph algorithms on the 2-section can be and often is lacking and that the research community needs to develop more algorithms that deal with hyper-graphs directly.

While non-uniform random hypergraphs might serve as the very first model of the real-world hypergraphs, the assumption that events that occur in the network are independent is likely not reasonable. Perhaps of particular impor-tance is a notion of clustering coefficient; there have been a number of proposals for generalizing clustering coefficient from graphs to hypergraphs, for instance [1,5,11,14,15]. In the longer journal version of this paper we compute the hyper-graph clustering coefficient of [15] for our random hypergraph model and the two real networks we are investigating. With the knowledge and experience we gath-ered, we feel that we are better prepared to propose a probabilistic model that is more suitable. However, it is left for the forthcoming papers.

Due to space limitation, all proofs and details of a set of experiments we performed in this project are omitted in this proceedings version but will be included in the journal version of this paper.

2   Definitions and Conventions

2.1   Random Graphs and Random Hypergraphs

First, let us recall a classic model of random graphs. The *binomial random graph* $\mathscr{G}(n,p)$ is the random graph G with vertex set $[n] := \{1, 2, \ldots, n\}$ in which every pair $\{i, j\} \in \binom{[n]}{2}$ appears independently as an edge in G with probability p. Note that $p = p(n)$ may (and usually does) tend to zero as n tends to infinity.

In this paper, we are concerned with more general combinatorial objects: hypergraphs. A *hypergraph* H is an ordered pair $H = (V, E)$, where V is a finite set (the *vertex set*) and E is a family of distinct subsets of V (the *hyperedge set*). A hypergraph $H = (V, E)$ is *r-uniform* if all hyperedges of H are of size r. For a given $r \in \mathbb{N}$, the *random r-uniform hypergraph* $\mathscr{H}_r(n, p)$ has n labelled vertices from a vertex set $V = [n]$, in which every subset $e \subseteq V$ of size $|e| = r$ is chosen to be a hyperedge of H randomly and independently with probability p. For $r = 2$, this model reduces to the model $\mathscr{G}(n, p)$.

The binomial random graph model is well known and thoroughly studied (e.g. [3,10,12]). Random hypergraphs are much less understood and, unfortunately, most of the existing papers deal with uniform hypergraphs. For example, Hamilton cycles (both tight ones and loose ones) were recently studied in [7–9]; perfect matchings were investigated in [13] (for a few more examples see the recent book on random graphs [10]).

In this paper, we are concerned with a natural generalization of this model that produces non-uniform hypergraphs. Let $\boldsymbol{p} = (p_r)_{r \geq 1}$ be any sequence of numbers such that $0 \leq p_r = p_r(n) \leq 1$ for each $r \geq 1$. The *random hypergraph* $\mathscr{H}(n, \boldsymbol{p})$ has n labelled vertices from a vertex set $V = [n]$, in which every subset $e \subseteq V$ of size $|e| = r$ is chosen to be a hyperedge of H randomly and independently with probability p_r. In other words, $\mathscr{H}(n, \boldsymbol{p}) = \bigcup_{r \geq 1} \mathscr{H}_r(n, p_r)$ is a union of independent uniform hypergraphs.

Let us mention that there are several natural generalizations that might be worth exploring, depending on a specific application in mind. One possible generalization would be to allow hyperedges to contain repeated vertices (multiset-hyperedge hypergraphs). Another one would be to allow the hyperedges to be chosen with possible repetitions, to get parallel hyperedges.

A vertex of a hypergraph is *isolated* if it is contained in no edge. (In particular, a vertex of degree 1 that belongs only to an edge of size 1 is not isolated.) The *2-section* of a hypergraph H, denoted $[H]_2$, is the graph on the same vertex set as H and an edge uv if (and only if) u and v are contained in some edge of H. In other words, it is obtained by making each hyperedge of H a clique in $[H]_2$.

2.2 Subgraphs

In this paper, we are concerned with occurrences of a given substructure in hypergraphs. However, there are at least two natural generalizations of "subgraph" for hypergraphs.

A hypergraph $H' = (V', E')$ is a *strong subhypergraph* (called *hypersubgraph* by Bahmanian and Sajna [2] and *partial hypergraph* by Duchet [6]) of $H = (V, E)$ if $V' \subseteq V$ and $E' \subseteq E$; that is, each hyperedge of H' is also an hyperedge of H. We write $H' \subseteq_s H$ when H' is a strong subhypergraph of H. For $H = (V, E)$ and $V' \subseteq V$, the *strong subhypergraph of H induced by V'*, denoted $H_s[V']$, has vertex set V' and hyperedge set $E' = \{e \in E : e \subseteq V'\}$.

The hypergraph H' is a *weak subhypergraph* of H (called *subhypergraph* by Bahmanian and Sajna) if $V' \subseteq V$ and $E' \subseteq \{e \cap V' : e \in E\}$; that is, each hyperedge of H' can be extended to one of H by adding vertices of $V \setminus V'$ to it.

For $V' \subseteq V$, the *weak subhypergraph induced by* V', denoted $H_w[V']$, has vertex set V' and hyperedge set $E' = \{e \cap V' : e \in E\}$. For this paper however, since we desire our hypergraphs to never contain the empty hyperedge, we tacitly replace E' by $E' \setminus \{\emptyset\}$. For now, weak subgraphs are assumed not to have multiple hyperedges (E' is a set, not a multiset).

Note that when G is an ordinary (i.e. 2-uniform) graph, strong subhypergraphs are the usual notion of subgraph, and weak subhypergraphs are subgraphs together with possible hyperedges of size 1. Note that each strong subhypergraph is also a weak subhypergraph but not vice versa.

Given hypergraphs H_1 and H_2, a weak (resp. strong) *copy* of H_1 in H_2 is a weak (resp. strong) subhypergraph of H_2 isomorphic to H_1. Most of this paper is concerned with determining the existence of strong or weak copies of a fixed H in $\mathscr{H}(n, \boldsymbol{p})$. With a mild abuse of terminology, we will often say that $\mathscr{H}$ contains H as a weak (strong) subhypergraph when we actually mean that $\mathscr{H}$ contains a weak (strong) copy of H. The precise meaning will always be clear from the context (Fig. 1).

$$H_1 \qquad\qquad H_2$$

Fig. 1. The hypergraph H_1 appears as a weak subhypergraph of H_2 (induced by the dashed vertex subset), but not as a strong subhypergraph.

3   Small Subgraphs in $\mathscr{H}(n, p)$

We are interested in answering questions about the existence of subgraphs within $\mathscr{H}(n, \boldsymbol{p})$. This question was addressed for $\mathscr{G}(n, p)$ by Bollobás in [4]. We are going to generalize his result to hypergraphs but first we need a few more definitions. Let $H = (V, E)$ be a hypergraph. Denote by $v(H) = |V|$ and by $e(H) = |E|$ the number of vertices and edges of H, respectively. For any $r \geq 1$, we will use $e_r(H) = |\{e \in E : |e| = r\}|$ to denote the number of edges of H of size r.

Define

$$\mu_s(H) = n^{v(H)} \prod_{r \geq 1} p_r^{e_r(H)}. \tag{1}$$

Now we are ready to state our result for the appearance of strong subgraphs of $\mathscr{H}(n, \boldsymbol{p})$. We adopt the convention that $0^0 = 1$ and assume all our hypergraphs have nonempty vertex set.

Theorem 1. *Let H be an arbitrary fixed hypergraph. Let $\boldsymbol{p} = (p_r)_{r \geq 1}$ be any sequence such that $0 \leq p_r = p_r(n) \leq 1$ for each $r \geq 1$. Let $\mathscr{J}$ denote the family of all strong subgraphs of H.*

(a) *If for some* $H' \in \mathcal{J}$ *we have* $\mu_s(H') \to 0$ *(as* $n \to \infty$*), then a.a.s.* $\mathcal{H}(n, \boldsymbol{p})$
    does not contain H *as a strong subgraph.*
(b) *If for all* $H' \in \mathcal{J}$ *we have* $\mu_s(H') \to \infty$ *(as* $n \to \infty$*), then a.a.s.* $\mathcal{H}(n, \boldsymbol{p})$
    contains H *as a strong subgraph.*

Let us mention that the result also holds for the multiset setting: that is, when vertices are allowed to be repeated in each hyperedge with some multiplicity. Moreover, if additionally there exists $\varepsilon > 0$ such that $p_r \leq 1 - \varepsilon$, for all r, then the same conditions (that is, conditions (a) and (b) of Theorem 1) determine whether or not H appears as an *induced* strong subgraph.

In view of Theorem 1, we emphasize that the existence of strong copies of H in $\mathcal{H}(n, \boldsymbol{p})$ cannot be determined by translating to graphs via 2-sections. For instance, consider the three hypergraphs H_1, H_2 and H_3 from Fig. 2. Each of these has H_1 as its 2-section. However, the expected number of strong copies of H_1, H_2 and H_3 in $\mathcal{H}(n, \boldsymbol{p})$ is, respectively, of order $n^4 p_2^5$, $n^4 p_2^2 p_3$, and $n^4 p_3^2$. So if, say, $p_3 = n^{-5/2}$ and $p_2 = n^{-3/4}$, then we expect many copies of H_1, a constant number of copies of H_2, and $o(1)$ copies of H_3. Moreover, by testing the conditions of Theorem 1 for all the strong subgraphs of H_1, H_2, H_3, we obtain that a.a.s. $\mathcal{H}(n, \boldsymbol{p})$ contains H_1 but not H_3 as a strong subgraph (and the theorem is inconclusive for H_2).

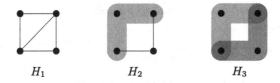

$$H_1 \qquad\qquad H_2 \qquad\qquad H_3$$

Fig. 2. These three hypergraphs have the same 2-section, which is precisely H_1, but their behaviour as potential strong subgraphs of $\mathcal{H}(n, \boldsymbol{p})$ is different.

Now we move to our result for the appearance of weak subgraphs of $\mathcal{H}(n, \boldsymbol{p})$. For technical reasons, we restrict ourselves to hypergraphs with bounded edge sizes. Formally, for a given $M \in \mathbb{N}$, we say that $H = (V, E)$ is an *M-bounded hypergraph* if $|e| \leq M$ for all $e \in E$. Similarly, $\boldsymbol{p} = (p_r)_{r \geq 1}$ is an *M-bounded sequence* if $p_r = 0$ for $r > M$. We will use $\boldsymbol{p} = (p_r)_{r=1}^{M}$ for an M-bounded sequence instead of an infinite sequence $\boldsymbol{p} = (p_r)_{r \geq 1}$ with a bounded number of non-zero values. Clearly, if $\boldsymbol{p}$ is M-bounded, then so is $\mathcal{H}(n, \boldsymbol{p})$ (with probability 1). For $r \in [M]$, let

$$p_r' = p_r + n p_{r+1} + n^2 p_{r+2} + \cdots + n^{M-r} p_M, \qquad (2)$$

and, given any fixed hypergraph H, define

$$\mu_w(H) = n^{v(H)} \prod_{r=1}^{M} (p_r')^{e_r(H)}, \qquad (3)$$

which will play an analogous role to $\mu_s(H)$.

Theorem 2. *Let H be an arbitrary fixed hypergraph, and let $\mathcal{J}$ be the collection of all strong subgraphs of H. Let $\mathbf{p} = (p_r)_{r=1}^{M}$ be an M-bounded sequence.*

(a) If for some $H' \in \mathcal{J}$ we have $\mu_w(H') \to 0$ (as $n \to \infty$), then a.a.s. $\mathscr{H}(n, \mathbf{p})$ does not contain H as a weak subgraph.

(b) If for all $H' \in \mathcal{J}$ we have $\mu_w(H') \to \infty$ (as $n \to \infty$), then a.a.s. $\mathscr{H}(n, \mathbf{p})$ contains H as a weak subgraph.

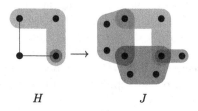

$$H \qquad\qquad\qquad J$$

Fig. 3. A hypergraph J and an induced weak hypergraph H with different thresholds for appearance as strong subgraphs.

We shall discuss a few relevant points concerning Theorem 2. First, it is possible that a.a.s. some graph occurs as a weak subgraph but not as a strong one. For example, if

$$p_1 = n^{-0.6}, \quad p_2 = n^{-0.9}, \quad p_3 = n^{-1.7}, \quad \text{and} \quad p_4 = n^{-3.1}, \tag{4}$$

then a.a.s. $\mathscr{H}(n, \mathbf{p})$ does not contain graph H (presented on Fig. 3) as a strong subgraph but a.a.s. it contains J (also presented on Fig. 3) and so a.a.s. it contains H as a weak subgraph.

Next, observe that if we replace $\mathcal{J}$ in the statement of Theorem 2 by the collection $\mathcal{J}_w$ of all *weak* subgraphs of H, the theorem remains valid. This is trivially true for part (b), since $\mathcal{J}_w \supseteq \mathcal{J}$. For part (a), a few easy modifications in the proof are necessary which will be mentioned in the journal version of this paper.

Finally, let us comment on the definition of p'_r, and introduce related parameters p''_r and p'''_r, which will play a role later on. Our particular choice of p'_r in (3) and thus in the statement of Theorem 2 is the simplest function from the equivalence class of all functions of the same order. However, the following one is more natural (as argued below). For $r \in [M]$, let

$$p''_r = p_r + n p_{r+1} + \binom{n}{2} p_{r+2} + \cdots + \binom{n}{M-r} p_M. \tag{5}$$

Note that p'_r and p''_r are of the same order. More precisely,

$$(1 + o(1)) \frac{p'_r}{(M-r)!} \leq p''_r \leq p'_r.$$

Hence, p'_r can be replaced in (3) by the more natural (but less simple) p''_r, and Theorem 2 remains valid. It is worth noting that both p'_r and p''_r can be greater than one or even tend to infinity as $n \to \infty$. Indeed, p''_r is not a probability but rather is asymptotic to the expected number of edges to which a given set of size r belongs. In contrast, the probability that such a set belongs to some edge is

$$p'''_r = 1 - (1 - p_r)(1 - p_{r+1})^{n-r}(1 - p_{r+2})^{\binom{n-r}{2}} \cdots (1 - p_M)^{\binom{n-r}{M-r}}. \qquad (6)$$

Observe that, if $p'_r = o(1)$ (or equivalently $p''_r = o(1)$), then $p''_r, p''_{r+1}, \ldots, p''_M = o(1)$, and therefore

$$p'''_r = 1 - \exp\left(-(1 + o(1))\left(p_r + np_{r+1} + \binom{n}{2}p_{r+2} + \cdots + \binom{n}{M-r}p_M\right)\right)$$

$$= 1 - \exp\left(-(1 + o(1))p''_r\right) \sim p''_r, \qquad (7)$$

so p''_r and p'''_r asymptotically coincide.

4   Induced Weak Subgraphs

Let us discuss how one can use Theorem 2 to determine whether H appears as an *induced* weak subgraph of $\mathscr{H}(n, \boldsymbol{p})$. This seems to be more complex than in the case of strong subgraphs: the non-edges of H play a crucial role in determining the existence of induced weak copies. Indeed, a weak subgraph H of $\mathscr{H}(n, \boldsymbol{p})$ is induced provided that, for every set e of vertices of H that do not form an edge, e cannot be extended to an edge of $\mathscr{H}(n, \boldsymbol{p})$ by adding vertices not in H.

First, we will give some conditions that forbid a.a.s. the existence of weak induced copies of H in $\mathscr{H}(n, \boldsymbol{p})$ (even if H *does* appear as a weak subgraph).

Proposition 1. *Let H be an arbitrary fixed hypergraph on k vertices with a non-edge of size r $(1 \le r \le k)$. Suppose $p''_r \ge (k + \varepsilon) \log n$ for some constant $\varepsilon > 0$. Then, a.a.s. H does not occur as an induced weak subgraph of $\mathscr{H}(n, \boldsymbol{p})$.*

As a result, the condition $p''_r \ge (k + \varepsilon) \log n$ implies that, if H is an induced weak subgraph of $\mathscr{H}(n, \boldsymbol{p})$ of order k, then H must contain all possible edges of size r. Coming back to our example with H from Fig. 3 and p_i's from (4), note that $p''_1 \sim \binom{n}{2}p_3 \sim n^{0.3}/2$. Thus, a.a.s. H will not occur as an induced weak subgraph of $\mathscr{H}(n, \boldsymbol{p})$, as not every vertex of H belongs to an edge of size 1.

On the other hand, suppose that $r \ge 1$ is the size of the smallest non-edge of H and assume that

$$\max\{p'''_r, p'''_{r+1}, \ldots, p'''_M\} \le 1 - \varepsilon \qquad (8)$$

for some constant $\varepsilon > 0$. Then any given weak copy of H in $\mathscr{H}(n, \boldsymbol{p})$ is also induced with probability bounded away from zero. In that case, the same calculations in the proof of Theorem 2 (that is omitted in this version) are still valid with an extra $\Theta(1)$ factor, and thus the conclusions of that theorem extend to induced weak subgraphs. Since verifying condition (8) may sometimes be slightly unwieldy, we will give a simpler sufficient condition.

Proposition 2. *Let H be an arbitrary fixed hypergraph, and let r be the size of its smallest non-edge. Suppose that $p_r \leq 1 - \varepsilon$ for some constant $\varepsilon > 0$ and that $p'_r = O(1)$ (and, as a result, $p''_r = O(1)$ too). If the conditions in part (b) of Theorem 2 are satisfied, then a.a.s. $\mathcal{H}(n, \boldsymbol{p})$ contains H as an induced weak subgraph.*

Let us come back to our example from Fig. 3 and (4) for the last time. Note that $p''_2 \sim np_3 = n^{-0.7} = o(1)$. Hence, if the "missing" edges of size 1 are added to H, then a.a.s. the resulting graph would occur as an induced weak subgraph of $\mathcal{H}(n, \boldsymbol{p})$.

5   The 2-section of $\mathcal{H}(n, \boldsymbol{p})$

We first consider the question of whether a given (2-uniform) graph G appears as a subgraph of the 2-section of $\mathcal{H}(n, \boldsymbol{p})$. We again may assume that G has no isolated vertices.

Let us start with some general observations that apply for any host hypergraph $\mathcal{H}$, not necessarily $\mathcal{H}(n, \boldsymbol{p})$. Observe that $G \subseteq [\mathcal{H}]_2$ if and only if there is a weak subhypergraph H of $\mathcal{H}$ such that G is a spanning subgraph of $[H]_2$. So we may test for $G \subseteq [\mathcal{H}]_2$ by finding every hypergraph H with G a spanning subgraph of $[H]_2$ and applying Theorem 2 to each. We can reduce the number of hypergraphs that need to be tested: if H_1 is a weak subhypergraph of H_2 and H_2 is a weak subhypergraph of $\mathcal{H}$, then H_1 is also a weak subhypergraph of $\mathcal{H}$. Note too that a spanning weak subhypergraph is actually a strong subhypergraph. So it suffices to check only the hypergraphs H that are minimal—with respect to the (strong) subhypergraph relation—that have G as a spanning subgraph of their 2-section.

In $\mathcal{H}(n, \boldsymbol{p})$, one can reduce the number of hypergraphs H to be tested even further. A *subedge system* of a hypergraph H is a hypergraph obtained from H by taking a subset of each edge of H and taking a (strong) subhypergraph of the result. Let H_1 be a subedge system of H_2 and let H_2 be a weak subhypergraph of H. It is not necessarily true that H_1 is a weak subhypergraph of H, but it is true a.a.s. for $H = \mathcal{H}(n, \boldsymbol{p})$.

Proposition 3. *Let H_1 and H_2 be fixed hypergraphs with H_1 a spanning subedge system of H_2, and let $\boldsymbol{p}$ be M-bounded. Let $\mathcal{J}_1$ and $\mathcal{J}_2$ denote the set of all strong subgraphs of H_1 and H_2, respectively. If every $H'_2 \in \mathcal{J}_2$ satisfies $\mu_w(H'_2) \to \infty$, then every $H'_1 \in \mathcal{J}_1$ also satisfies $\mu_w(H'_1) \to \infty$.*

Corollary 1. *Fix a (2-uniform) graph G without isolated vertices. Let $\mathcal{F}$ denote the family of minimal—with respect to the subedge system relation—hypergraphs containing G in their 2-section. Let $\boldsymbol{p}$ be M-bounded.*

(a) If for every $H \in \mathcal{F}$ there is some strong subgraph $H' \subseteq_s H$ with $\mu_w(H') \to 0$, then a.a.s. G is not a subgraph of $[\mathcal{H}(n, \boldsymbol{p})]_2$.

(b) If for some $H \in \mathcal{F}$ every strong subgraph $H' \subseteq_s H$ satisfies $\mu_w(H') \to \infty$, then a.a.s. G is a subgraph of $[\mathcal{H}(n, \boldsymbol{p})]_2$.

We next consider the following problem. Suppose that a copy of G is found in $[\mathcal{H}(n, \boldsymbol{p})]_2$. We would like to estimate the probability that this copy comes from a given weak subhypergraph of $\mathcal{H}(n, \boldsymbol{p})$.

Let G be a fixed 2-uniform graph with no isolated vertices. Let $\mathcal{F}$ denote the family of hypergraphs H on the same vertex set as G such that $G \simeq [H]_2$. Then, G appears as an induced subgraph of $[\mathcal{H}(n, \boldsymbol{p})]_2$ if and only if some $H \in \mathcal{F}$ appears as an induced weak subhypergraph of $\mathcal{H}(n, \boldsymbol{p})$. More precisely, for every set of vertices S inducing a copy of G in $[\mathcal{H}(n, \boldsymbol{p})]_2$, there is exactly one $H \in \mathcal{F}$ such that S induces a weak copy of H in $\mathcal{H}(n, \boldsymbol{p})$. We say in that case that hypergraph H *originates* that particular copy of G. As a result we have the following proposition.

Proposition 4. *Let $\boldsymbol{p} = (p_r)_{r=1}^M$ be an M-bounded sequence. For $r \in [M]$, let p_r''' be defined as in* (6). *Then, given a copy of G in $[\mathcal{H}(n, \boldsymbol{p})]_2$, the probability that it originates from a given $H \in \mathcal{F}$ is*

$$(1 + o(1))\frac{\mathrm{aut}(H) \prod_{r=1}^M (p_r''')^{e_r(H)}(1 - p_r''')^{\binom{v(G)}{r} - e_r(H)}}{\sum_{H' \in \mathcal{F}} \mathrm{aut}(H') \prod_{r=1}^M (p_r''')^{e_r(H')}(1 - p_r''')^{\binom{v(G)}{r} - e_r(H')}}.$$

Instead of determining which specific $H \in \mathcal{F}$ originates a copy of G in the 2-section of $\mathcal{H}(n, \boldsymbol{p})$, we may take equivalence classes in $\mathcal{F}$ given their r-edge counts. To that end, define the *signature* of $H \in \mathcal{F}$ as the vector $\mathbf{e}(H) = (e_1(H), e_2(H), \ldots, e_k(H))$, where $k = v(G)$ (and hence also $k = v(H)$). Let $\mathbf{e}(\mathcal{F}) = \{\mathbf{e}(H) : H \in \mathcal{F}\}$. For a given signature $\mathbf{e} \in \mathbf{e}(\mathcal{F})$, let $\mathcal{F}_{\mathbf{e}} \subseteq \mathcal{F}$ be the family of hypergraphs in $\mathcal{F}$ with signature $\mathbf{e}$. Notice that $\{\mathcal{F}_{\mathbf{e}} : \mathbf{e} \in \mathbf{e}(\mathcal{F})\}$ is a partition of $\mathcal{F}$. Then, the following useful result holds.

Corollary 2. *Let $\boldsymbol{p} = (p_r)_{r=1}^M$ be an M-bounded sequence. For $r \in [M]$, let p_r''' be defined as in* (6). *Then, given a copy of G in $[\mathcal{H}(n, \boldsymbol{p})]_2$, the probability that it originates from a hypergraph with a given signature $\mathbf{e} = (m_1, m_2, \ldots, m_k) \in \mathbf{e}(\mathcal{F})$ is*

$$(1 + o(1))\frac{\sum_{H \in \mathcal{F}_{\mathbf{e}}} \mathrm{aut}(H) \prod_{r=1}^k (p_r''')^{m_r}(1 - p_r''')^{\binom{v(G)}{r} - m_r}}{\sum_{H' \in \mathcal{F}} \mathrm{aut}(H') \prod_{r=1}^k (p_r''')^{e_r(H')}(1 - p_r''')^{\binom{v(G)}{r} - e_r(H')}}.$$

The following example illustrates how, under natural assumptions on $\boldsymbol{p}$, Corollary 2 implies that a copy of G in $[\mathcal{H}(n, \boldsymbol{p})]_2$ "typically" originates from a hypergraph $H \in \mathcal{F}$ with few but large edges rather than many but small edges. Let $G = K_k$ (i.e. the clique of order k) for a fixed $k \geq 2$, and suppose that $\boldsymbol{p}$ is an M-bounded sequence satisfying $\binom{n}{j}p_j = O(n)$ for all $j \in [M]$. The latter condition is equivalent to assuming that the expected number of edges of each given size is at most linear in the number of vertices, which is a fairly reasonable assumption for many relevant models of hypergraph networks. Suppose additionally that for some r with $k \leq r \leq M$ we also have $\binom{n}{r}p_r = \Omega(n)$. From (7), we obtain that $p_j''' = O(1/n^{j-1})$ for every $j \in [M]$ and $p_k''' = \Theta(1/n^{k-1})$. Consider the signature $\hat{\mathbf{e}} = (0, \ldots, 0, 1)$ corresponding to the hypergraph $\hat{H}$ on k

vertices with one single edge of size k. A straightforward inductive argument reveals that, for any signature $\mathbf{e} = (m_1, m_2, \ldots, m_k) \in \mathbf{e}(\mathcal{F})$,

$$\prod_{r=1}^{k} (p_r''')^{m_r} (1 - p_r''')^{\binom{k}{r} - m_r} = \begin{cases} (1 + o(1)) p_k''' = \Theta(1/n^{k-1}) & \text{if } \mathbf{e} = \widehat{\mathbf{e}} \\ o(1/n^{k-1}) & \text{if } \mathbf{e} \neq \widehat{\mathbf{e}}. \end{cases}$$

As a result, applying Corollary 2 to all signatures different from $\widehat{\mathbf{e}}$, we conclude that, for a given copy of G in $[\mathscr{H}(n, \boldsymbol{p})]_2$, a.a.s. it must originate from $\widehat{H}$.

6   Experiments

We performed a number of experiments on two real-world datasets that are naturally represented as a hypergraph network. Our goal was to compare the results with the corresponding theoretical predictions. Due to space limitation, the details are omitted in this version but will be included in the journal version of this paper.

The experiments we performed confirmed the intuition that the fact that some set of vertices S forms a hyperedge should increase the probability that some proper subset of S belongs to some other hyperedge. Moreover, in many instances, the correlation seems to be so strong that not only having one hyperedge increases substantially the probability that another hyperedge intersects it but it is more likely that there will be another hyperedge intersecting it than not. Of course, such behaviour is not present in our theoretical model in which events are independent. In order to understand the behaviour we experience, some notion of "clustering coefficient" has to be introduced in the hypergraph setting. Again, the details are omitted here but will be included in the journal version of this paper.

7   Conclusions and Future Work

The goal of the larger project behind this paper is to propose a reasonable model for complex networks using hypergraphs, as they seem more suitable for many existing networks and associated applications. Whereas there are many models using graphs (classic ones such as $\mathscr{G}(n, p)$, random d-regular graphs, and PA model, as well as spatial ones such as random geometric graphs and SPA model), there are very few using hypergraphs. In order to better understand micro-processes that shape macro-properties that are observed in these networks, we introduced the random hypergraphs and investigated some properties of it in order to compare them with two real-world networks. These results are interesting from a pure random graph theory perspective but, of course, we did not expect such models to work well in practice; we did it to learn why they do *not* work. As is common in this field, such an exercise taught us a lot, and we feel that we are now better prepared to design a more suitable model, probably combining both geometry and the "rich get richer" paradigm. However, it is left for the forthcoming papers.

References

1. Alexander, M., Robins, G.: Small worlds among interlocking directors: network structures and distance in bipartite graphs. Comput. Math. Org. Theor. **10**(1), 69–94 (2004)
2. Bahmanian, M., Sajna, M.: Connection, separation in hypergraphs (2015). arXiv:1504.04274v1
3. Bollobás, B.: Random Graphs. Cambridge University Press, Cambridge (2001)
4. Bollobás, B.: Random graphs. In: Temperley, H.N.V. (ed.) Combinatorics. London Mathematical Society Lecture Note Series, vol. 52, pp. 80–102. Cambridge University Press, Cambridge (1981)
5. Borgatti, S., Everett, M.: Network analysis of 2-mode data. Soc. Netw. **19**(3), 243–269 (1997)
6. Duchet, P.: Hypergraphs. In: Graham, R.L., Grötschel, M., Lovász, L. (eds.) Handbook of Combinatorics. Elsevier, Amsterdam (1995)
7. Dudek, A., Frieze, A.M.: Loose Hamilton cycles in random k-uniform hypergraphs. Electron. J. Comb. **17**, P48 (2011)
8. Dudek, A., Frieze, A.M.: Tight Hamilton cycles in random uniform hypergraphs. Random Struct. Algorithms **42**, 374–385 (2012)
9. Ferber, A.: Closing gaps in problems related to Hamilton cycles in random graphs and hypergraphs (preprint)
10. Frieze, A.M., Karoński, M.: Introduction to Random Graphs. Cambridge University Press, Cambridge (2015)
11. Le Blond, S., Guillaume, J.-L., Latapy, M.: Clustering in P2P exchanges and consequences on performances. In: Castro, M., Renesse, R. (eds.) IPTPS 2005. LNCS, vol. 3640, pp. 193–204. Springer, Heidelberg (2005). doi:10.1007/11558989_18
12. Janson, S., Łuczak, T., Ruciński, A.: Random Graphs. Wiley, New York (2000)
13. Johansson, A., Kahn, J., Vu, V.: Factor in random graphs. Random Struct. Algorithms **33**, 1–28 (2008)
14. Latapy, M., Magnien, C., Del Vecchio, N.: Basic notions for the analysis of large two-mode networks. Soc. Netw. **30**(1), 31–48 (2008)
15. Zhou, W., Nakhleh, L.: Properties of metabolic graphs: biological organization or representation artifacts? BMC Bioinform. **12**, 132 (2011)

The Spread of Cooperative Strategies on Grids with Random Asynchronous Updating

Christopher Duffy$^{(\boxtimes)}$ and Jeannette Janssen

Department of Mathematics and Statistics, Dalhousie University, Halifax, Canada
`christopher.duffy@dal.ca`

Abstract. The Prisoner's Dilemma Process on a graph is an iterative process where each vertex, with a fixed strategy (*cooperate* or *defect*), plays the game with each of its neighbours. At the end of a round each vertex may change its strategy to that of its neighbour with the highest pay-off. Here we study the spread of cooperative and selfish behaviours on a toroidal grid, where each vertex is initially a cooperator with probability p. When vertices are permitted to change their strategies via a randomized asynchronous update scheme, we find that for some values of p the limiting ratio of cooperators may be modelled as a polynomial in p. Theoretical bounds for this ratio are confirmed via simulation.

1 Introduction and Preliminaries

The particular topology of a network has a dramatic impact on discrete processes that model competitive interactions in communities [6]. For example, spread of a particular attitude or belief is less likely to propagate completely in Erdős-Renyi graphs, than on small-world networks [2]. Studies of cellular automata indicate that the particular updating scheme impacts the limiting configuration of randomly seeded cellular automaton [8]. Here we combine these two paradigms to study a discrete-time process that may be modelled as a cellular automaton with a particular updating scheme.

The Prisoner's Dilemma, a staple of classical game theory, is a 2-player game in which each of the two players simultaneously make a decision to either *cooperate* or *defect*. Each of the players receives a pay-off whose amount takes into account the decisions of both players. Classically the game is played in a single round. However by considering the game as being played in a series of rounds, the Prisoner's Dilemma may be used to model a variety of scenarios in many disciplines, including evolutionary biology [3], economics [5] and sociology [9].

Here we consider the iterated Prisoner's Dilemma as a game played between neighbours on a graph. In each round each vertex plays, with a fixed strategy (*cooperate* or *defect*), the game with each of its neighbours. The score for each vertex is the sum of the pay-offs its receives in each game. At the end of each round, vertices are given the opportunity to update their strategy to that of their most successful neighbour. A survey of the literature in this area reveals a multitude of experimental results on a variety of graphs with a variety of

© Springer International Publishing AG 2016
A. Bonato et al. (Eds.): WAW 2016, LNCS 10088, pp. 152–163, 2016.
DOI: 10.1007/978-3-319-49787-7_13

update schemes [7,11]. Here we discuss an updating scheme that considers the set of vertices envious of their neighbours and updates them in a random order, playing a round of the game after each individual vertex has updated. This update scheme behaves similarly to the random independent model for updating cellular automata [8]; however, it provides necessary structure to facilitate proofs of observed behaviours. As such we provide theoretical results for the survival rates of cooperators in toroidal grids seeded with random initial strategies. More broadly, the Prisoner's Dilemma on graphs fits in the context of evolutionary games on graphs. A survey of methods and research in this area is given in [10].

Let $G = (V, E)$ be a graph. A *configuration*, C, is function that assigns a strategy to each vertex of G. Formally, $C : V \rightarrow \{0, 1\}$, where 0 corresponds to *defector* and 1 corresponds to *cooperator*. The *pay-off function*, f, assigns the score for the first player to an ordered pair that represents the strategies of the first and second player. The pay-off function $f : \{0, 1\} \times \{0, 1\} \rightarrow \{0, 1, T\}$ is given by

$C(v_1), C(v_2)$	$f(C(v_1), C(v_2))$
$(0, 0)$	0
$(1, 0)$	0
$(0, 1)$	T
$(1, 1)$	1

where $T > 1$ is a fixed constant. We refer to T as the *cheating advantage*.

Let C be fixed and let $v \in V(G)$. The *score of v with respect to C* is given by

$$s(v) = \sum_{x \in N(v)} f(C(v), C(x)).$$

When the context is clear we refer to the *score of v*. The *most successful neighbours of v* are the vertices in the closed neighbourhood of v (denoted $N[v]$) that have the greatest score. We may choose T so that each of the most successful neighbours have the same strategy. Let u be a most successful neighbour of v. The vertex v is called *weak with respect to C* if $C(v) \neq C(u)$. Otherwise, we say that v is *strong*. We are interested in the change in the configuration with respect to time; we use C_t to denote the configuration at time t and s_t to denote the score at time t.

The configuration, D, resulting from *updating v with respect to C* is given by

$$D(x) = \begin{cases} C(x), & x \neq v, \text{ or} \\ C(x), & x \text{ is strong, or} \\ 1 - C(x), & x \text{ is weak.} \end{cases}$$

Note that when we update v with respect to C it is only the strategy of v and the scores in its closed neighbourhood that possibly change.

We call a maximal connected proper subgraph of vertices with the same strategy a *k-cluster*. We say that a 1-cluster is an *isolated* vertex. The *k*-cluster $H \leq G$ has a *border of width b* if for all $h \in V(H)$ and all $v \in V(G - H)$ such that $C(h) = C(v)$, then $d(h, v) \geq b$.

Given a graph G and some $T > 1$, our process is initialised with C_0, some configuration of the vertices. The process proceeds as follows. Let W_t be the set of weak vertices with respect to C_t. If $W_t = \emptyset$, then the process terminates. In this case we say that C_t is a *stable configuration*. Otherwise, we select with uniform probability a permutation, σ, of the elements of W_t. Considering the permutation as a sequence of the elements of W_t, we proceed through $|W_t|$ subrounds. At the kth subround we update the kth vertex of the sequence, v_k, with respect to the current configuration (i,e., the configuration resulting from the $(k - 1)^{th}$ subround). We refer to the process as *the Prisoner's Dilemma process on G with randomised asynchronous updating*. Though, for brevity we refer to this process as the *PD process on G*.

As we are interested in the spread of the cooperative strategy, for any configuration we may consider the ratio of cooperators. For configuration C_t, let r_t be the ratio of cooperators to $|V(G)|$. If C_t is *stable configuration*, then we define the *final ratio*, denoted r_f, to be r_t.

Consider the following example on the 6×6 grid to highlight how the choice of T for the process and the updating permutation in a particular round affect the spread of strategies. Consider the configuration given in Fig. 1a. In our figures we use white squares for cooperators and grey squares for defectors.

If $T = \frac{5}{3}$ each of the cooperators have score 2 and each of the labelled defectors have score $\frac{5}{3}$. All unlabelled vertices have score 0, as they are defectors with no cooperator neighbours. Observe that $W_0 = \{v_1, v_2, \ldots, v_8\}$. Figure 1b gives the resulting configurations after applying $\sigma_1 = (v_2, v_3, v_6, v_7, v_1, v_4, v_5, v_8)$ to C_0 and alternatively applying $\sigma_2 = (v_2, v_4, v_6, v_8, v_1, v_3, v_5, v_7)$ to C_0. In the first case, after subround 4, v_1 is no longer a weak vertex, and so does not change strategy. The resulting configuration has no weak vertices. However, in the second case v_1 is a weak vertex after subround 4, and so does change from being a cooperator to a defector. Here the resulting configuration has 8 weak vertices.

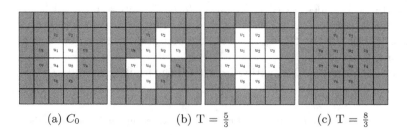

(a) C_0 (b) $T = \frac{5}{3}$ (c) $T = \frac{8}{3}$

Fig. 1. The PD Process on the 6×6 grid with varying values of T.

For $T = \frac{8}{3}$, each of the cooperators have score 2, and each of the labelled defectors have score $\frac{8}{3}$. All unlabelled vertices have score 0, as they are defectors with no cooperator neighbours. Observe that $W_0 = \{u_1, u_2, u_3, u_4\}$. Regardless of the choice σ, Fig. 1c is the resulting configuration after round 0.

A configuration C_t is called *forced* if C_{t+1} will be the configuration regardless of the choice of σ at time t. For a sequence $C_0, C_1, \ldots$ of configurations, we call the sequence resulting from removing the forced configurations and re-indexing the *forced sequence*. We use the notation $C'_0, C'_1, \ldots$ to refer a forced sequence.

Our example shows that value of T influences the spread of strategies. In a 4-regular graph if $1 < T < \frac{4}{3}$, then most successful neighbour of v is the vertex in the closed neighbourhood of v with the most cooperator neighbours, with defectors taking precedence in the case of a draw. For the remainder of this paper we consider only the case $1 < T < \frac{4}{3}$ as we restrict our study to the toroidal grid. We use the notation $T = 1 + \epsilon$ to refer to T in this range. We say that a defector with k cooperator neighbours has score $k + \epsilon$.

In this paper we study the resulting behaviour of the PD process on toroidal grids where for each vertex of the grid, v, $C_0(v) = 1$ with probability $p \in [0, 1]$. Figure 2 gives examples of starting and the resulting stable configurations for various values of p. Here we notice that though the initial configuration is randomised, the resulting stable configuration exhibits a surprising amount of structure. Though the update process introduces uncertainty through the choice of the permutation of the weak vertices, we find that for some values of p, we may predict r_t as $t \to \infty$. In Sect. 2 we consider the growth of small clusters existing in infinite grids. We use the results from Sect. 2 in Sect. 3 to derive probabilistic

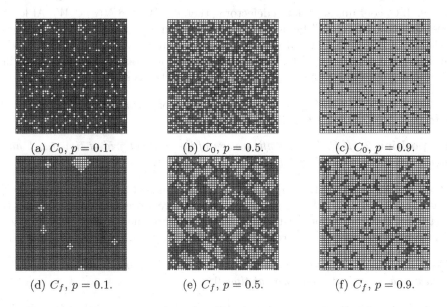

(a) C_0, $p = 0.1$.   (b) C_0, $p = 0.5$.   (c) C_0, $p = 0.9$.

(d) C_f, $p = 0.1$.   (e) C_f, $p = 0.5$.   (f) C_f, $p = 0.9$.

Fig. 2. Initial and resulting configurations on the 50×50 toroidal grid for random initial configurations and varying values of p.

bounds on r_f on the $n \times n$ toroidal grid for a fixed value of p, and for p as a function of n.

2   Evolution of Clusters of Cooperators and Defectors in Infinite Grids

Observe that a k-cluster situated in a grid is a fixed polyomino of order k. For $k > 0$ let α_k be the number of polyominoes of order k. From [1] we get the following initial terms for the sequence $\{\alpha_k\}_{k \geq 1}$: $1, 2, 6, 19, 63 \ldots$. In this section we consider the evolution of small clusters situated in infinite grids. We apply these results in Sect. 3 to study the behaviour of r_t in the $n \times n$ toroidal grid.

Up to rotation and reflection of the infinite grid, there is a single 1-cluster of defectors and a single 2-cluster of defectors. A configuration of a single defector in an infinite grid of cooperators has exactly four weak vertices – the neighbours of the single defector. By examining the number of cooperator neighbours, we see that when one of these weak vertices has changed to a cooperator the resulting configuration is stable. Therefore a 1-cluster of defectors in a field of cooperators evolves to a 2-cluster of defectors, which is a stable configuration. To show that the spread of a k-cluster of defectors in an infinite grid of cooperators is bounded, we require the following results.

Proposition 1. *If C_t is the configuration of the toroidal grid at time t, then for all $t > 0$ there is no isolated defector in C_t.*

Proof. If C_t contains an isolated defector v, then $v \notin W_t$ and $N(v) \subset W_t$. At least one element of $N(v)$ will be a defector in C_{t+1} and so v will not be an isolated defector in C_{t+1}. No isolated defector can be created in the tth round, as such an isolated defector would be a cooperator with four cooperator neighbours in subround in which it changes. However, such a vertex is strong.   □

For a configuration C_i, we say that a cooperator v is a *persistent cooperator after time i* if $C_t(v) = 1$ for all $t \geq i$.

Proposition 2. *If C_t is the configuration of the toroidal grid at time $t > 1$ and $s_t(v) = 4$, then v is a persistent cooperator at time t.*

Proof. By Proposition 1, $s_t(u) \leq 4$ for all $t > 0$ and all $u \in V$, as cooperators have score at most 4 and non-isolated defectors have score no more than $3 + \epsilon$. If $s_t(v) = 4$, then v is a cooperator with four cooperator neighbours. This implies that each vertex of $N[v]$ is strong. Therefore if $s_t(v) = 4$, then $s_{t+1}(v) = 4$.   □

Corollary 3. *If $s_0(v) = 4$ and v has no isolated defector at distance 2 in C_0, then v is a persistent cooperator at time 0.*

If v is a persistent cooperator at time 0, then we say that v is a *initial persistent cooperator*.

Corollary 4. *If C_0 is the configuration of the infinite grid consisting of a k-cluster of defectors in a field of cooperators such that the k-cluster is contained within an rectangle of length ℓ and width w ($\ell, w \in \mathbb{N}$), then there exists a rectangle of length $\ell + 4$ and width $w + 4$ so that the growth of the defector strategy is contained within this rectangle.*

Proof. Every cooperator at distance 2 from the cluster of defectors is an initial persistent cooperator. □

As we consider the evolution of k-clusters of cooperators in a field of defectors we encounter some cases for which there are no surviving cooperators. In this case, we say that the particular cluster evolves to an *empty cluster*.

Up to rotation and reflection, there is a single 1-cluster and a single 2-cluster. When placed in a sufficiently large grid of defectors, each of these clusters evolve to an empty cluster after at most two time steps.

Up to rotation and reflection there are two species of 3-clusters: 3-*lines* and 3-*corners*. When placed in a sufficiently field of defectors 3-lines evolve to a stable configuration containing a 5-cluster with probability 1. 3-corners evolve to a stable configuration containing a 5-cluster with probability $\frac{1}{2}$ and to an empty cluster with probability $\frac{1}{2}$. The evolution of these clusters is given in Fig. 3a. Note that weak vertices are indicated with a circle.

Up to rotation and reflection there are 5 species of 4-clusters: 4-lines, 4-corners, 4-hats, 4-turns, and 4-squares (See Fig. 3). A 4-hat stabilises to a stable 5-cluster with probability 1. However, for each of the other configurations there is non-zero probability of large growth. The evolution of these clusters through a small number of iterations is given in Fig. 3.

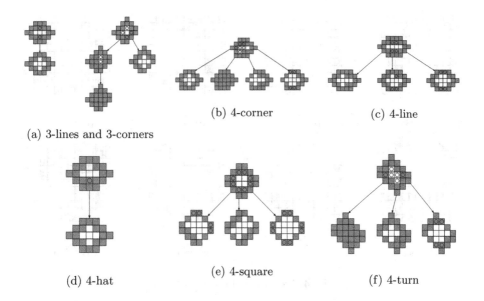

(a) 3-lines and 3-corners

(b) 4-corner

(c) 4-line

(d) 4-hat

(e) 4-square

(f) 4-turn

Fig. 3. Evolution of 3-clusters and 4-clusters.

We wish to show that a 4-cluster in an infinite grid of defectors will eventually evolve to a stable configuration. Though the growth such clusters pass through many different configurations, we show that every configuration in the sequence of forced configurations where C_0 is a 4-cluster of cooperators an infinite field of defectors can be classified in to one of 8 types. We consider these types equivalent under rotation and reflection. The width parameter of each type tells the number of columns that contain collaborators, whereas the height parameter tells us how many collaborators are in the column with the greatest amount of collaborators.

A *stable cluster of width w and height h* ($h \equiv 1 \mod 2$) consists of cooperators in w columns. The first column (starting from the left) contains a single cooperator. The number of cooperators increase by 2 in each subsequent column, until the maximum h is reached. After the maximum is reached, columns of height h may repeat an arbitrary number of times. The heights of the columns then decrease by 2 in each subsequent column until there is a column with a single cooperator. An example is given in Fig. 4a. Observe that such a configuration is stable when placed in a field of defectors. We use the symbol $\diamondsuit_{w,h}$ to denote a stable cluster of width w and height h.

A *square cluster of height h and width w* ($h \equiv 0 \mod 2$) is a cluster consisting of cooperators in w columns. The first column cooperator contains two cooperators. The number of cooperators increase by 2 in each subsequent column, until the maximum h is reached. After the maximum is reached, columns of height h may repeat an arbitrary number of times. The heights of the columns then decrease by 2 in each subsequent column until there is a column with a two cooperators. An example is given in Fig. 4b. We use the symbol $\diamondsuit_{w,h}$ to denote

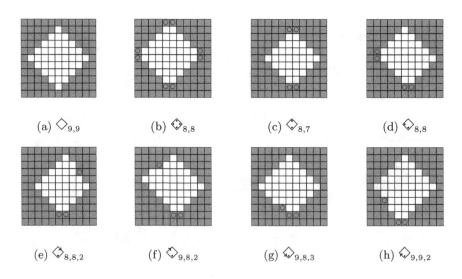

(a) $\diamondsuit_{9,9}$ (b) $\diamondsuit_{8,8}$ (c) $\diamondsuit_{8,7}$ (d) $\diamondsuit_{8,8}$

(e) $\diamondsuit_{8,8,2}$ (f) $\diamondsuit_{9,8,2}$ (g) $\diamondsuit_{9,8,3}$ (h) $\diamondsuit_{9,9,2}$

Fig. 4. Examples of configurations

a square cluster of height h and width w. Observe that $\diamondsuit_{w,h}$ has 8 weak vertices when placed in a field of defectors.

A 2-*opposite cluster of height h and width w ($h \equiv 1 \bmod 2$)* is a cluster of cooperators in w columns. The first column cooperator contains one cooperator. The number of cooperators increases by 2 in each subsequent column, until h is reached. This is followed by at least one column containing h cooperators. The heights of the columns then decrease by 2 in each subsequent column until there is a column with a single cooperator. An example is given in Fig. 4c. We use the symbol $\diamondsuit_{w,h}$ to denote an 2-opposite cluster of height h and width w. Observe that $\diamondsuit_{w,h}$ has 4 weak vertices when placed in a field of defectors.

A 2-*adjacent cluster of height h and width w ($h \equiv 0 \bmod 2$)* is a cluster of cooperators in w columns. The first column cooperator contains two cooperators. The number of cooperators increase by 2 in each subsequent column, until $h-1$ is reached. This is followed by an arbitrary number of columns of height h. The heights of the columns then decrease by 2 in each subsequent column until there is a column with a single cooperator. An example is given in Fig. 4d. We use the symbol $\diamondsuit_{w,h}$ to denote an 2-adjacent cluster of height h and width w. Observe that $\diamondsuit_{w,h}$ has 4 weak vertices when placed in a field of defectors.

A 1-*opposite cluster of width w, height h and length ℓ ($\ell < h/2$)* is formed from a 2-opposite cluster of height h and width w by changing one of the four weak defectors to be cooperators when $\ell = 1$; or formed from a 1-opposite cluster of height h and width w and length $\ell - 1$ by changing the single weak defector with no weak defector neighbours to be a cooperator when $\ell \geq 2$. We use the symbol $\diamondsuit_{w,h,\ell}$ to denote a 1-opposite cluster of width w, height h and length ℓ. See Fig. 4e for an example. Observe that $\diamondsuit_{w,h,\ell}$ has 3 weak vertices when placed in a field of defectors.

A 1-*adjacent cluster of width w, height h and length ℓ of type A ($\ell < h/2$)* is formed from a 2-adjacent cluster of height h and width w by changing the lower of the two leftmost weak defectors to be a cooperator when $\ell = 1$; or formed from a 1-adjacent cluster of width w, height h and length $\ell - 1$ of type A by changing the single weak defector with no weak defector neighbours to be a cooperator when $\ell \geq 2$. We use the symbol $\diamondsuit_{w,h,\ell}$ to denote a 1-adjacent cluster of width w, height h and length ℓ of type A. See Fig. 4g for an example. Observe that $\diamondsuit_{w,h,\ell}$ has 3 weak vertices when placed in a field of defectors.

A 1-*adjacent cluster of width w, height h and length ℓ of type B ($\ell < h/2$)* is formed from a 2-adjacent cluster of height h and width w by changing the upper of the two left most weak defectors to be a cooperator when $\ell = 1$; or formed from a 1-adjacent cluster of width w, height h and length $\ell - 1$ of type B by changing the single weak defector with no weak defector neighbours to be a cooperator when $\ell \geq 2$. We use the symbol $\diamondsuit_{w,h,\ell}$ to denote a 1-adjacent cluster

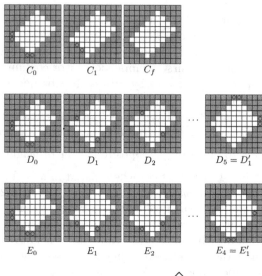

Fig. 5. Evolution of $\diamondsuit_{8,7}$

of width w, height h and length ℓ of type B. See Fig. 4f for an example. Observe that $\diamondsuit_{w,h,\ell}$ has 3 weak vertices when placed in a field of defectors.

A 1-*adjacent cluster of width* w, *height* h *and length* ℓ *of type C* ($\ell < h/2$) is a cluster consisting of cooperators in w columns. When $\ell = 1$, the first column cooperator contains a single cooperator. The number of cooperators increase by 2 in each subsequent column, until $h - 1$ is reached. This is followed by one column with h cooperators. The amount of cooperators then decrease by 2 in subsequent columns, until there is a column with a single cooperator. When $\ell \geq 2$, it is formed from a 1-adjacent cluster of width w, height h and length $\ell - 1$ of type C by changing the single weak defector with no weak defector neighbours to be a cooperator. We use the symbol $\diamondsuit_{w,h,\ell}$ to denote a 1-adjacent cluster of width w, height h and length ℓ of type C. See Fig. 4h for an example. Observe that $\diamondsuit_{w,h,\ell}$ has 3 weak vertices when placed in a field of defectors.

Let $\mathcal{D}$ be the set of these configurations. Consider the sequences of configurations given in Fig. 5. Though $C_0 = D_0 = E_0 = \diamondsuit_{8,7}$, the choice of σ gives three distinct possibilities for the following configuration. Up to symmetry, C_1 and D_1 each occur with probability $\frac{1}{4}$, and E_1 occurs with probability $\frac{1}{2}$. After proceeding through the forced iterations, we see that in each case we arrive at an element of $\mathcal{D}$. In particular, $\diamondsuit_{8,7}$ transitions to $\diamondsuit$ with probability $\frac{1}{4}$, to $\diamondsuit_{8,9}$ with probability $\frac{1}{4}$ and to $\diamondsuit_{9,8,3}$ with probability $\frac{1}{2}$.

By examining the other cases when $C_i' \in \mathcal{D}$, we may observe that if $C_i' \in \mathcal{D}$, then $C_{i+1}' \in \mathcal{D}$. Figure 6 gives the transitions that do not depend in the w and h.

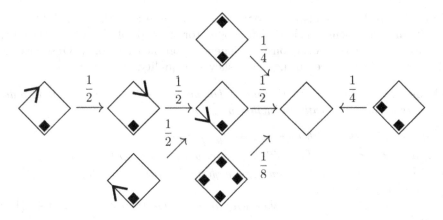

Fig. 6. Transition diagram of $\mathcal{D}$ for transitions that do not depend on h and w.

Lemma 5. *If $C_i' \in \mathcal{D}$, then with probability at most $\frac{1}{8}$ either $C_{i+1}' = \lozenge$ or $C_{i+2}' = \lozenge$ or $C_{i+3}' = \lozenge$.*

Proposition 6. *If C_0 is a 4-cluster of cooperators in an infinite grid of defectors, then the PD process terminates with probability 1.*

Proof. If C_0 is not an element of $\mathcal{D}$, then $C_1 \in \mathcal{D}$ with $w = 4$ and $h = 3$. The result now follows from Lemma 5. □

3   Growth in the Toroidal Grid

We now consider the behaviour of r_t for various regimes of p on the $n \times n$ toroidal grid. In particular we examine two cases. Firstly, we consider p to be a fixed constant. Then we consider p as a function of n.

Theorem 7. *Consider the PD process on the $n \times n$ toroidal grid where $C_0(v) = 1$ with fixed probability $p \in (0, 1)$ for all $v \in V(G)$. With high probability $r_t > p^{13}$ for all $t \geq 0$.*

Proof. By Corollary 3, if there exists at least m initial persistent cooperators, then $r_t > \frac{m}{n^2}$ for all $t \geq 0$. Let p' be the probability that a vertex is an initial persistent cooperator. We proceed using Chebyshev's inequality, letting $\epsilon = p' - p^{13}$.

Lemma 8. *Consider the PD process on an $n \times n$ toroidal grid where $C_0(v) = 1$ with probability $p = f(n)$. Let k be a positive integer and K be a k-cluster of cooperators. If $p \gg n^{-\frac{2}{k}}$ and $p(n) \to 0$ as $n \to \infty$, then the number of copies of K in C_0 is $n^2 p^k (1 + o(1))$ with high probability.*

Proof. Let K be a k-cluster of cooperators. Observe that the probability of a particular vertex being the lower left cooperator of a copy of K in C_0 is $p^k(1-p)^c$, where c is a positive integer constant depending on the shape of K. Observe that $(1-p)^c \sim 1$. We proceed using Chebyshev's inequality, letting $\epsilon = q^{\frac{1}{4}}$.

Theorem 9. *Consider the Prisoner's Dilemma process on an $n \times n$ toroidal grid where $C_0(v) = 1$ with probability $p = f(n)$.*

1. *If $p \ll n^{-\frac{2}{3}}$, then with high probability $r_f = 0$;*
2. *if $n^{-\frac{2}{3}} \ll p \ll n^{-\frac{1}{2}}$, then with high probability $r_f = 20p^3(1+o(1))$;*
3. *if $n^{-\frac{2}{3}} \ll p \ll n^{-\frac{2}{5}}$, then with high probability $r_f \leq 20p^3(1+o(1)) + 19\log^4(n)p^4(1+o(1))$;*
4. *if $1 - n^{-1} \ll p \ll 1 - n^{-2}$, then with high probability $r_f = (2p-1)(1+o(1))$; and*
5. *if $1 - n^{-2} \ll p$, then with high probability $r_f = 1$.*

Proof. Note that if $p \ll n^{-\frac{2}{k}}$ then with high probability there are no k'-clusters of cooperators for any $k' \geq k$. By applying Markov's inequality we may show that any cluster of cooperators (defectors) in C_0 has a border of width sufficient to contain its growth. As such we may treat each cluster as if it is situated in an infinite field of defectors (cooperators). Using the results from Sect. 2 and Lemma 8 we find the given bounds for r_f.

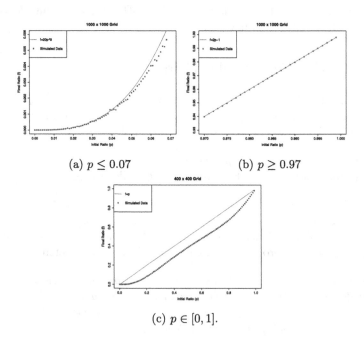

(a) $p \leq 0.07$　　　　　　　(b) $p \geq 0.97$

(c) $p \in [0,1]$.

Fig. 7. Experimental results

3.1   Results of Simulation

Fig. 7 gives plots of the results of simulation on the 1000×1000 toroidal grid and the 400×400 toroidal grid. For the 1000×1000 toroidal grid we simulate the PD process for $p \in \{0.001, 0.002, \dots, 0.07\}$ and $p \in \{0.920, 0.921, \dots, 0.999\}$. For the 400×400 toroidal grid we simulate the PD process for $p \in \{0.01, 0.02, \dots, 0.99\}$. For each graph and each value of p, there were 10 simulations. The data points shown give the mean of r_f over the 10 simulations. The results agree with the conclusions drawn in Theorem 9. On the 400×400 toroidal grid, for very small values of p the curve appears to be cubic, and for very large values of p the curve appears to be linear, as expected. It appears that other regimes do exist. However exploring these regimes further using similar methods would require greater computing resources. Code, datasets and full details available at [4].

References

1. On-line encyclopaedia of integer sequences https://oeis.org, sequence A001168
2. Choi, H., Kim, S.H., Lee, J.: Role of network structure and network effects in diffusion of innovations. Ind. Mark. Manage. **39**(1), 170–177 (2010)
3. Weerd, H., Verbrugge, R.: Evolution of altruistic punishment in heterogeneous populations. J. Theor. Biol. **290**, 88–103 (2011)
4. Duffy, C., Janssen, J.: The prisoners dilemma on toroidal grids (2016). http://www.mathstat.dal.ca/cduffy/PD, electronic resource
5. Gibbons, R.: Game theory for applied economists. Princeton University Press, Princeton (1992)
6. Nicosia, V., Bagnoli, F., Latora, V.: Impact of network structure on a model of diffusion and competitive interaction. EPL (Europhys. Lett.) **94**(6), 68009 (2011)
7. Nowak, M.A., May, R.M.: The spatial dilemmas of evolution. Int. J. Bifurcat. Chaos **3**(01), 35–78 (1993)
8. Schönfisch, B., Roos, A.: Synchronous and asynchronous updating in cellular automata. BioSystems **51**(3), 123–143 (1999)
9. Sigmund, K., Nowak, M.: Cooperation in heterogeneous populations. In: Fischer, G.H., Laming, D. (eds.) Contributions to Mathematical Psychology, Psychometrics, and Methodology. Recent Research in Psychology, pp. 223–235. Springer, Heidelberg (1994)
10. Szabó, G., Fáth, G.: Evolutionary games on graphs. Phys. Rep. **446**(46), 97–216 (2007)
11. Szabó, G., Vukov, J., Szolnoki, A.: Phase diagrams for an evolutionary prisoners dilemma game on two-dimensional lattices. Phys. Rev. E **72**(4), 047107 (2005)

Author Index

Printed in the United States
By Bookmasters